MULTIMEDIA FUNDAMENTALS VOLUME 1:

Media Coding and Content Processing

IMSC Press Multimedia Series

ANDREW TESCHER, Series Editor, *Compression Science Corporation*

Advisory Editors

LEONARDO CHIARIGLIONE, *CSELT*
TARIQ S. DURRANI, *University of Strathclyde*
JEFF GRALNICK, *E-splosion Consulting, LLC*
CHRYSOSTOMOS L. "MAX" NIKIAS, *University of Southern California*
ADAM C. POWELL III, *The Freedom Forum*

▶ Desktop Digital Video Production
   Frederic Jones

▶ Touch in Virtual Environments:
   Haptics and the Design of Interactive Systems
   Edited by Margaret L. McLaughlin,
   João P. Hespanha, and Gaurav S. Sukhatme

▶ The MPEG-4 Book
   Edited by Fernando M. B. Pereira and Touradj Ebrahimi

▶ Multimedia Fundamentals, Volume 1:
   Media Coding and Content Processing
   Ralf Steinmetz and Klara Nahrstedt

▶ Intelligent Systems for Video Analysis and Access Over the Internet
   Wensheng Zhou and C. C. Jay Kuo

The Integrated Media Systems Center (IMSC), a National Science Foundation Engineering Research Center in the University of Southern California's School of Engineering, is a preeminent multimedia and Internet research center. IMSC seeks to develop integrated media systems that dramatically transform the way we work, communicate, learn, teach, and entertain. In an integrated media system, advanced media technologies combine, deliver, and transform information in the form of images, video, audio, animation, graphics, text, and haptics (touch-related technologies). IMSC Press, in partnership with Prentice Hall, publishes cutting-edge research on multimedia and Internet topics. IMSC Press is part of IMSC's educational outreach program.

Library of Congress Cataloging-in-Publication Data

Steinmetz, Ralf.
　Multimedia fundamentals / Ralf Steinmetz, Klara Nahrstedt
　　p.　cm.
　Includes bibliographical references and index.
　Contents: v. 1. Media coding and content processing.
　ISBN 0-13-031399-8
　1. Multimedia systems. I. Nahrstedt, Klara II. Title.

QA76.575 .S74 2002
006.7—dc21　　　　　　　　　　　　　　　　2001056583
　　　　　　　　　　　　　　　　　　　　　　　　CIP

Editorial/Production Supervision: *Nick Radhuber*
Acquisitions Editor: *Bernard Goodwin*
Editorial Assistant: *Michelle Vincente*
Marketing Manager: *Dan DePasquale*
Manufacturing Buyer: *Alexis Heydt-Long*
Cover Design: *John Christiana*
Cover Design Director: *Jerry Votta*

© 2002 by Prentice Hall PTR
Prentice-Hall, Inc.
Upper Saddle River, NJ 07458

Prentice Hall books are widely used by corporations and government agencies for training, marketing, and resale.

The publisher offers discounts on this book when ordered in bulk quantities. For more information, contact Corporate Sales Department, phone: 800-382-3419; fax: 201-236-7141; email: corpsales@prenhall.com Or write: Corporate Sales Department, Prentice Hall PTR, One Lake Street, Upper Saddle River, NJ 07458.

Product and company names mentioned herein are the trademarks or registered trademarks of their respective owners.

All rights reserved. No part of this book may be reproduced, in any form or by any means, without permission in writing from the publisher.

Printed in the United States of America

10 9 8 7 6 5 4 3 2

ISBN 0-13-031399-8

Pearson Education LTD.
Pearson Education Australia PTY, Limited
Pearson Education Singapore, Pte. Ltd
Pearson Education North Asia Ltd
Pearson Education Canada, Ltd.
Pearson Educación de Mexico, S.A. de C.V.
Pearson Education—Japan
Pearson Education Malaysia, Pte. Ltd
Pearson Education, Upper Saddle River, New Jersey

Multimedia Fundamentals Volume 1:

Media Coding and Content Processing

Ralf Steinmetz
Klara Nahrstedt

PRENTICE HALL PTR
UPPER SADDLE RIVER, NJ 07458
WWW.PHPTR.COM

Contents

Preface   xv

1 Introduction   1
   1.1 Interdisciplinary Aspects of Multimedia 2
   1.2 Contents of This Book 3
   1.3 Organization of This Book 4
        1.3.1 Media Characteristics and Coding 5
        1.3.2 Media Compression 5
        1.3.3 Optical Storage 6
        1.3.4 Content Processing 6
   1.4 Further Reading About Multimedia 6

2 Media and Data Streams   7
   2.1 The Term "Multimedia" 7
   2.2 The Term "Media" 7
        2.2.1 Perception Media 8
        2.2.2 Representation Media 8
        2.2.3 Presentation Media 8
        2.2.4 Storage Media 9
        2.2.5 Transmission Media 9

		2.2.6	Information Exchange Media 9

        2.2.6   Information Exchange Media 9
        2.2.7   Presentation Spaces and Presentation Values 9
        2.2.8   Presentation Dimensions 10
  2.3   Key Properties of a Multimedia System 11
        2.3.1   Discrete and Continuous Media 12
        2.3.2   Independent Media 12
        2.3.3   Computer-Controlled Systems 12
        2.3.4   Integration 12
        2.3.5   Summary 13
  2.4   Characterizing Data Streams 13
        2.4.1   Asynchronous Transmission Mode 13
        2.4.2   Synchronous Transmission Mode 14
        2.4.3   Isochronous Transmission Mode 14
  2.5   Characterizing Continuous Media Data Streams 15
        2.5.1   Strongly and Weakly Periodic Data Streams 15
        2.5.2   Variation of the Data Volume of Consecutive Information Units 16
        2.5.3   Interrelationship of Consecutive Packets 18
  2.6   Information Units..19

3 Audio Technology   21

  3.1   What Is Sound? 21
        3.1.1   Frequency 22
        3.1.2   Amplitude 23
        3.1.3   Sound Perception and Psychoacoustics 23
  3.2   Audio Representation on Computers 26
        3.2.1   Sampling Rate 27
        3.2.2   Quantization 27
  3.3   Three-Dimensional Sound Projection 28
        3.3.1   Spatial Sound 28
        3.3.2   Reflection Systems 30
  3.4   Music and the MIDI Standard 30
        3.4.1   Introduction to MIDI 31
        3.4.2   MIDI Devices 31
        3.4.3   The MIDI and SMPTE Timing Standards 32
  3.5   Speech Signals 32

        3.5.1   Human Speech 32
        3.5.2   Speech Synthesis 33
  3.6   Speech Output 33
        3.6.1   Reproducible Speech Playout 34
        3.6.2   Sound Concatenation in the Time Range 34
        3.6.3   Sound Concatenation in the Frequency Range 36
        3.6.4   Speech Synthesis 36
  3.7   Speech Input 37
        3.7.1   Speech Recognition 38
  3.8   Speech Transmission 40
        3.8.1   Pulse Code Modulation 40
        3.8.2   Source Encoding 41
        3.8.3   Recognition-Synthesis Methods 42
        3.8.4   Achievable Quality 43

4 Graphics and Images   45

  4.1   Introduction 45
  4.2   Capturing Graphics and Images 46
        4.2.1   Capturing Real-World Images 46
        4.2.2   Image Formats 48
        4.2.3   Creating Graphics 53
        4.2.4   Storing Graphics 54
  4.3   Computer-Assisted Graphics and Image Processing 55
        4.3.1   Image Analysis 56
        4.3.2   Image Synthesis 71
  4.4   Reconstructing Images 72
        4.4.1   The Radon Transform 73
        4.4.2   Stereoscopy 74
  4.5   Graphics and Image Output Options 75
        4.5.1   Dithering 76
  4.6   Summary and Outlook 77

5 Video Technology   79

  5.1   Basics 79

		5.1.1	Representation of Video Signals	79

      5.1.1 Representation of Video Signals 79
      5.1.2 Signal Formats 83
  5.2 Television Systems 87
      5.2.1 Conventional Systems 87
      5.2.2 High-Definition Television (HDTV) 88
  5.3 Digitization of Video Signals 90
      5.3.1 Composite Coding 91
      5.3.2 Component Coding 91
  5.4 Digital Television 93

6 Computer-Based Animation   95

  6.1 Basic Concepts 95
      6.1.1 Input Process 95
      6.1.2 Composition Stage 96
      6.1.3 Inbetween Process 96
      6.1.4 Changing Colors 97
  6.2 Specification of Animations 97
  6.3 Methods of Controlling Animation 98
      6.3.1 Explicitly Declared Control 98
      6.3.2 Procedural Control 99
      6.3.3 Constraint-Based Control 99
      6.3.4 Control by Analyzing Live Action 99
      6.3.5 Kinematic and Dynamic Control 100
  6.4 Display of Animation 100
  6.5 Transmission of Animation 101
  6.6 Virtual Reality Modeling Language (VRML) 101

7 Data Compression   105

  7.1 Storage Space 105
  7.2 Coding Requirements 106
  7.3 Source, Entropy, and Hybrid Coding 110
      7.3.1 Entropy Coding 110
      7.3.2 Source Coding 111
      7.3.3 Major Steps of Data Compression 111

7.4	Basic Compression Techniques		113
	7.4.1	Run-Length Coding	113
	7.4.2	Zero Suppression	113
	7.4.3	Vector Quantization	114
	7.4.4	Pattern Substitution	114
	7.4.5	Diatomic Encoding	114
	7.4.6	Statistical Coding	114
	7.4.7	Huffman Coding	115
	7.4.8	Arithmetic Coding	116
	7.4.9	Transformation Coding	117
	7.4.10	Subband Coding	117
	7.4.11	Prediction or Relative Coding	117
	7.4.12	Delta Modulation	118
	7.4.13	Adaptive Compression Techniques	118
	7.4.14	Other Basic Techniques	120
7.5	JPEG		120
	7.5.1	Image Preparation	122
	7.5.2	Lossy Sequential DCT-Based Mode	126
	7.5.3	Expanded Lossy DCT-Based Mode	132
	7.5.4	Lossless Mode	134
	7.5.5	Hierarchical Mode	135
7.6	H.261 (p×64) and H.263		135
	7.6.1	Image Preparation	137
	7.6.2	Coding Algorithms	137
	7.6.3	Data Stream	139
	7.6.4	H.263+ and H.263L	139
7.7	MPEG		139
	7.7.1	Video Encoding	140
	7.7.2	Audio Coding	144
	7.7.3	Data Stream	146
	7.7.4	MPEG-2	148
	7.7.5	MPEG-4	152
	7.7.6	MPEG-7	165
7.8	Fractal Compression		165
7.9	Conclusions		166

8 Optical Storage Media 169

- 8.1 History of Optical Storage 170
- 8.2 Basic Technology 171
- 8.3 Video Discs and Other WORMs 173
- 8.4 Compact Disc Digital Audio 175
  - 8.4.1 Technical Basics 175
  - 8.4.2 Eight-to-Fourteen Modulation 176
  - 8.4.3 Error Handling 177
  - 8.4.4 Frames, Tracks, Areas, and Blocks of a CD-DA 178
  - 8.4.5 Advantages of Digital CD-DA Technology 180
- 8.5 Compact Disc Read Only Memory 180
  - 8.5.1 Blocks 181
  - 8.5.2 Modes 182
  - 8.5.3 Logical File Format 183
  - 8.5.4 Limitations of CD-ROM Technology 184
- 8.6 CD-ROM Extended Architecture 185
  - 8.6.1 Form 1 and Form 2 186
  - 8.6.2 Compressed Data of Different Media 187
- 8.7 Further CD-ROM-Based Developments 188
  - 8.7.1 Compact Disc Interactive 188
  - 8.7.2 Compact Disc Interactive Ready Format 190
  - 8.7.3 Compact Disc Bridge Disc 191
  - 8.7.4 Photo Compact Disc 192
  - 8.7.5 Digital Video Interactive and Commodore Dynamic Total Vision .. 193
- 8.8 Compact Disc Recordable 194
- 8.9 Compact Disc Magneto-Optical 196
- 8.10 Compact Disc Read/Write 197
- 8.11 Digital Versatile Disc 198
  - 8.11.1 DVD Standards 198
  - 8.11.2 DVD-Video: Decoder 201
  - 8.11.3 Eight-to-Fourteen+ Modulation (EFM+) 201
  - 8.11.4 Logical File Format 202
  - 8.11.5 DVD-CD Comparison 202
- 8.12 Closing Observations 203

9 Content Analysis — 205

- 9.1 Simple vs. Complex Features 206
- 9.2 Analysis of Individual Images 207
  - 9.2.1 Text Recognition 207
  - 9.2.2 Similarity-Based Searches in Image Databases 209
- 9.3 Analysis of Image Sequences 210
  - 9.3.1 Motion Vectors 210
  - 9.3.2 Cut Detection 214
  - 9.3.3 Analysis of Shots 220
  - 9.3.4 Similarity-Based Search at the Shot Level 221
  - 9.3.5 Similarity-Based Search at the Scene and Video Level 224
- 9.4 Audio Analysis 226
  - 9.4.1 Syntactic Audio Indicators 226
  - 9.4.2 Semantic Audio Indicators 227
- 9.5 Applications 229
  - 9.5.1 Genre Recognition 229
  - 9.5.2 Text Recognition in Videos 233
- 9.6 Closing Remarks 234

Bibliography — 235

Index — 257

Preface

Multimedia Systems are becoming an integral part of our heterogeneous computing and communication environment. We have seen an explosive growth of multimedia computing, communication, and applications over the last decade. The World Wide Web, conferencing, digital entertainment, and other widely used applications are using not only text and images but also video, audio, and other continuous media. In the future, all computers and networks will include multimedia devices. They will also require corresponding processing and communication support to provide appropriate services for multimedia applications in a seamless and often also ubiquitous way.

This book is the first of three volumes that will together present the fundamentals of multimedia in a balanced way, particularly the areas of devices, systems, services and applications. In this book, we emphasize the field of multimedia devices. We also discuss how media data affects content processing. In Chapter 2 we present generic multimedia characteristics and basic requirements of multimedia systems. Chapters 3 through 6 discuss basic concepts of individual media. Chapter 3 describes audio concepts, such as sound perception and psychoacoustic, audio representation on computers; music and the MIDI standard; as well as speech signals with their input, output, and transmission issues. Chapter 4 concentrates on graphics and image characteristics, presenting image formats, image analysis, image synthesis, reconstruction of images as well as graphics and image output options. Chapter 5 goes into some detail about video signals, television formats, and digitization of video signals. Chapter 6 completes the presentation on individual media, addressing computer-based animation, its basic concepts, specification of animations, and methods of controlling them. Chapter 7 extensively describes compression concepts, such as run-length coding, Huffman coding, subband coding, and current compression standards such as JPEG, diverse MPEG formats, H.263 and others. Multimedia requires new considerations for storage devices, and we present in Chapter 8 basic optical storage technology as well as techniques that represent the core of the Compact Disc-Digital Audio (CD-DA), Compact Disc-Read Only Memory (CD-ROM), and Digital Versatile Disc (DVD) technologies. In Chapter 9, we summarize our conclusios utilizing the concepts in previous chapters and showing our projections for future needs in content processing and analysis.

Volume 1 will be followed by Volume 2 and Volume 3 of Multimedia Fundamentals. Volume 2 will concentrate on the operating system and networking aspects of distributed multimedia systems. Multimedia fundamentals in the System and Service domain will be covered, such as Quality of Service, soft-real-time scheduling, media servers and disk scheduling, streaming protocols, group communication, and synchronization. Volume 3 will emphasize some of the problems in the Service and Application domains of a distributed multimedia system. Coverage will include fundamental algorithms, concepts and basic principles in multimedia databases, multimedia programming, multimedia security, hypermedia documents, multimedia design, user interfaces, multimedia education, and generic multimedia applications for multimedia preparation, integration, transmission, and usage.

Overall the book has the character of a reference book, covering a wide scope. It has evolved from the third edition of our multimedia technology book, published in German in 2000 [Ste00]. (Figures from this book were reused with the permission of Springer-Verlag). However, several sections in the three upcoming volumes have changed from the corresponding material in the previous book. The results, presented in this book, can serve as a groundwork for the development of fundamental components at the device and storage levels in a multimedia system. The book can be used by computer professionals who are interested in multimedia systems or as a textbook for introductory multimedia courses in computer science and related disciplines. Throughout we emphasize how the handling of multimedia in the device domain will have clear implications in content processing.

To help instructors using this book, additional material is available via our Web site at *http://www.kom.e-technik.tu-darmstadt.de/mm-book/*. Please use *mm_book* and *mm_docs* for user name and password, respectively.

Many people have helped us with the preparation of this book. We would especially like to thank I. Rimac as well as M. Farber and K. Schork-Jakobi.

Last but not least, we would like to thank our families for their support, love, and patience.

CHAPTER 1

Introduction

Multimedia is probably one of the most overused terms of the 90s (for example, see [Sch97b]). The field is at the crossroads of several major industries: computing, telecommunications, publishing, consumer audio-video electronics, and television/movie/broadcasting. Multimedia not only brings new industrial players to the game, but adds a new dimension to the potential market. For example, while computer networking was essentially targeting a professional market, multimedia embraces both the commercial and the consumer segments. Thus, the telecommunications market involved is not only that of professional or industrial networks—such as medium- or high-speed leased circuits or corporate data networks—but also includes standard telephony or low-speed ISDN. Similarly, not only the segment of professional audio-video is concerned, but also the consumer audio-video market, and the associated TV, movie, and broadcasting sectors.

As a result, it is no surprise when discussing and establishing multimedia as a discipline to find difficulties in avoiding fuzziness in scope, multiplicity of definitions, and non-stabilized terminology. When most people refer to multimedia, they generally mean the combination of two or more continuous media, that is, media that have to be played during some well-defined time interval, usually with some user interaction. In practice, the two media are normally audio and video, that is, sound plus moving pictures.

One of the first and best known institutes that studied multimedia was the Massachusetts Institute of Technology (MIT) Media Lab in Boston, Massachusetts. MIT has been conducting research work in a wide variety of innovative applications, including personalized newspapers, life-sized holograms, or telephones that chat with callers

[Bra87]. Today, many universities, large-scale research institutes, and industrial organizations work on multimedia projects.

From the user's perspective, "multimedia" means that information can be represented in the form of audio signals or moving pictures. For example, movement sequences in sports events [Per97] or an ornithological lexicon can be illustrated much better with multimedia compared to text and still images only, because it can represent the topics in a more natural way.

Integrating all of these media in a computer allows the use of existing computing power to represent information interactively. Then this data can be transmitted over computer networks. The results have implications in the areas of information distribution and cooperative work. Multimedia enables a wide range of new applications, many of which are still in the experimental phase. Think for a moment that the World Wide Web (WWW) took its current form only at the beginning of the 90s. On the other hand, social implications inherent in global communication should not be overlooked. When analyzing such a broad field as multimedia from a scientific angle, it is difficult to avoid reflections on the effects of these new technologies on society as a whole. However, the sociological implications of multimedia are not the subject of this book. We are essentially interested in the technical aspects of multimedia.

1.1 Interdisciplinary Aspects of Multimedia

If we look at applications and technologies, there is a strong interest in existing multimedia systems and their constant enhancement. The process of change that takes place in the background in various industrial sectors should not be underestimated:

- The telecommunications industry used to be interested primarily in telephony. Today, telephone networks evolve increasingly into digital networks that are very similar to computer networks. Switching systems used to be made up of mechanical rotary switches. Today, they are computers. Conventional telephones have been evolving into computers, or they even exist as pure software in the form of "IP telephony."

- The consumer electronics industry—with its "brown ware"—contributed considerably to bringing down the price of video technology that is used in computers. Optical storage technology, for example, emerged from the success of CD players. Today, many manufacturers produce CD drives for computers and hi-fi equipment or television sets and computer screens.

- The TV and radio broadcasting sector has been a pioneer in professional audio-video technology. Professional systems for digital cutting of TV movies are commercially available today. Some of these systems are simple standard computers equipped with special add-on boards. Broadcasters now transmit their information

over cables so it is only natural that they will continue to become information vendors over computer networks in the future.
- Most publishing companies offer publications in electronic form. In addition, many are closely related to movie companies. These two industries have become increasingly active as vendors of multimedia information.

This short list shows that various industries merge to form interdisciplinary vendors of multimedia information.

Many hardware and software components in computers have to be properly modified, expanded, or replaced to support multimedia applications. Considering that the performance of processors increases constantly, storage media have sufficient capacities, and communication systems offer increasingly better quality, the overall functionality shifts more and more from hardware to software. From a technical viewpoint, the time restrictions in data processing imposed on all components represent one of the most important challenges. Real-time systems are expected to work within well-defined time limits to form fault-tolerant systems, while conventional data processing attempts to do its job as fast as possible.

For multimedia applications, fault tolerance and speed are not the most critical aspects because they use both conventional media and audio-video media. The data of both media classes needs to get from the source to the destination as fast as possible, i.e., within a well-defined time limit. However, in contrast to real-time systems and conventional data processing, the elements of a multimedia application are not independent from one another. In other words they do not only have to be integrated, they also have to be synchronized. This means that in addition to being an integrated system, composed of various components from both data types, there has to be some form of synchronization between these media.

Our goal is to present the multimedia systems from an integrated and global perspective. However, as outlined above, multimedia systems include many areas, hence we have decided to split the content about multimedia system fundamentals into three volumes. The first volume deals with media coding and content processing. The second volume describes media processing and communication. The third volume presents topics such as multimedia documents, security, and various applications.

1.2 Contents of This Book

If the word multimedia can have several meanings, there is a risk that the reader might not find what he or she is looking for. As mentioned above, this book is an integral part of a three-volume work on "Multimedia Fundamentals." Let us start by defining the scope of this first volume.

The primary objective of the book is to provide a comprehensive panorama of topics in the area of multimedia coding and content processing. It is structured as a reference book to provide fast familiarization with all the issues concerned. However, this book can also be used as a textbook for introductory multimedia courses. Many sections of this book explain the close relationships of the wide range of components that make up multimedia coding, compression, optical storage, and content processing in a multimedia system.

1.3 Organization of This Book

The overall goal is to present a comprehensive and practical view of multimedia technologies. Multimedia fundamentals can be divided as shown in Figure 1-1. We will present material about the most important multimedia fields in these volumes. The overall organization attempts to explain the largest dependencies between the components involved in terms of space and time. We distinguish between:

- *Basics*: In addition to the computer architecture for multimedia systems, one of the most important aspects is a media-specific consideration.
- *Systems*: This section covers system aspects relating to processing, storage, and communication and the relevant interfaces.
- *Services*: This section details single functions, which are normally implemented through individual system components.
- *Usage*: This section studies the type and design of applications and the interface between users and computer systems.

In this volume, we will present the basics of multimedia, concentrating on media-specific considerations such as individual media characteristics and media compression, and their dependencies on optical storage, content analysis, and processing.

Techniques like the sampling theorem or Pulse Code Modulation (PCM), discussed in a later chapter, with their respective mathematical background and practical implementations form the basis for digital audio-video data processing. Several techniques have evolved from these basics, each specialized for a specific medium. Audio technology includes music and voice processing. The understanding of video technology is essentially based on the development of digital TV technology, involving single pictures and animation. As demand on quality and availability of technologies increases, these media have high data rates, so that appropriate compression methods are necessary. Such methods can be implemented both by hardware and software. Furthermore, the high demand on quality and availability of multimedia technology has also placed heavy demands on optical storage systems to satisfy their requirements. As storage capacity and availability of other resources increase, content analysis is becoming an integral part of our multimedia systems and applications.

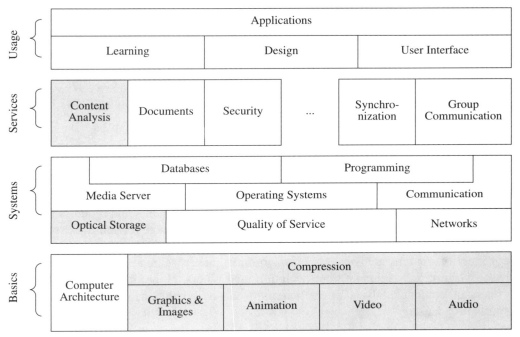

Figure 1-1 The most important multimedia fields, as discussed in this book.

1.3.1 Media Characteristics and Coding

The section on media characteristics and coding will cover areas such as sound characteristics with discussion of music and the MIDI standard, speech recognition and transmission, graphics and image coding characteristics. It will also include presentation of image processing methods and a video technology overview with particular emphasis on new TV formats such as HDTV. In addition to the basic multimedia data such as audio, graphics, images, and video, we present basic concepts for animation data and its handling within multimedia systems.

1.3.2 Media Compression

As the demand for high-quality multimedia systems increases, the amount of media to be processed, stored, and communicated increases. To reduce the amount of data, compression techniques for multimedia data are necessary. We present basic concepts of entropy and source compression techniques such as Huffman Coding or Delta Modulation, as well as hybrid video compression techniques such as MPEG-4 or H.263. In addition to the basic concepts of video compression, we discuss image and audio compression algorithms, which are of great importance to multimedia systems.

1.3.3 Optical Storage

Optical storage media offer much higher storage density at lower cost than traditional secondary storage media. We will describe various successful technologies that are the successors of long-playing records, such as audio compact discs and digital audio tapes. Understanding the basic concepts such as pits and lands in the substrate layers, modulation and error handling on CD-DA, and modes on CD-ROM, is necessary in order to understand the needs of media servers, disk management, and other components in multimedia systems.

1.3.4 Content Processing

Coding and storage directly or indirectly influence the processing and analysis of multimedia content in various documents. In recent years, due to the World Wide Web, we are experiencing wide distribution of multimedia documents and requests for multimedia information filtering tools using text recognition, image recognition, speech recognition, and other multimedia analysis algorithms. This section presents basic concepts of content analysis, such as similarity-based search algorithms, algorithms based on motion vectors and cut detection, and others. Hopefully this will clarify the effects of content analysis in applications such as television, movies, newscasts, or sports broadcasts.

1.4 Further Reading About Multimedia

Several fields discussed in this book are covered by other books in more detail. For example, multimedia databases are covered in [Mey91], video coding in [Gha99] and in the *Handbook of Multimedia Computing* [Fur98]. Moreover, the basics of audio technology, video technology, image processing, and various network systems are discussed in specialized papers and books, while this book describes all coding components involved in the context of integrated multimedia systems.

There is extensive literature on all aspects of multimedia. Some journals that frequently publish papers in this area are *IEEE Multimedia*, *IEEE Transaction on Multimedia*, *Multimedia Systems* (ACM Springer), and *Multimedia Tools and Applications*. Many other journals also publish papers on the subject.

In addition to a large number of national and international workshops in this field, there are several interdisciplinary, international conferences on multimedia systems, in particular: the ACM Multimedia Conference (the first conference took place in Anaheim, California, in August 1993), the IEEE Multimedia Conference (first conference held in May 1994), and the European Workshop on Interactive Distributed Multimedia Systems and Telecommunication Services (IDMS).

Media and Data Streams

This chapter provides an introduction to the terminology used in the entire book. We begin with our definition of the term multimedia as a basis for a discussion of media and key properties of multimedia systems. Next, we will explain data streams and information units used in such systems.

2.1 The Term "Multimedia"

The word multimedia is composed of two parts: the prefix *multi* and the root *media*. The prefix *multi* does not pose any difficulty; it comes from the latin word *multus*, which means "numerous." The use of *multi* as a prefix is not recent and many Latin words employ it.

The root *media* has a more complicated story. *Media* is the plural form of the Latin word *medium*. *Medium* is a noun and means "middle, center."

Today, the term multimedia is often used as an attribute for many systems, components, products, and concepts that do not meet the key properties we will introduce later (see Section 2.3). This means that the definition introduced in this book is (intentionally) restrictive in several aspects.

2.2 The Term "Media"

As with most generic words, the meaning of the word media varies with the context in which it is used. Our definition of medium is "a means to distribute and represent information." Media are, for example, text, graphics, pictures, voice, sound, and music. In this sense, we could just as well add water and the atmosphere to this definition.

[IHE93] provides a subtle differentiation of various aspects of this term by use of criteria to distinguish between perception, representation, presentation, storage, transmission, and information exchange media. The following sections describe these attributes.

2.2.1 Perception Media

Perception media refers to the nature of information perceived by humans, which is not strictly identical to the sense that is stimulated. For example, a still image and a movie convey information of a different nature, though stimulating the same sense. The question to ask here is: *How do humans perceive information?*

In this context, we distinguish primarily between what we see and what we hear. Auditory media include music, sound, and voice. Visual media include text, graphics, and still and moving pictures. This differentiation can be further refined. For example, a visual medium can consist of moving pictures, animation, and text. In turn, moving pictures normally consist of a series of scenes that, in turn, are composed of single pictures.

2.2.2 Representation Media

The term representation media refers to how information is represented internally to the computer. The encoding used is of essential importance. The question to ask here is: *How is information encoded in the computer?* There are several options:

- Each character of a piece of text is encoded in ASCII.
- A picture is encoded by the CEPT or CAPTAIN standard, or the GKS graphics standard can serve as a basis.
- An audio data stream is available in simple PCM encoding and a linear quantization of 16 bits per sampling value.
- A single image is encoded as Group-3 facsimile or in JPEG format.
- A combined audio-video sequence is stored in the computer in various TV standards (e.g., PAL, SECAM, or NTSC), in the CCIR-601 standard, or in MPEG format.

2.2.3 Presentation Media

The term presentation media refers to the physical means used by systems to reproduce information for humans. For example, a TV set uses a cathode-ray tube and loudspeaker. The question to ask here is: *Which medium is used to output information from the computer or input in the computer?*

We distinguish primarily between output and input. Media such as paper, computer monitors, and loudspeakers are output media, while keyboards, cameras, and microphones are input media.

2.2.4 Storage Media

The term storage media is often used in computing to refer to various physical means for storing computer data, such as magnetic tapes, magnetic disks, or digital optical disks. However, data storage is not limited to the components available in a computer, which means that paper is also considered a storage medium. The question to ask here is: *Where is information stored?*

2.2.5 Transmission Media

The term transmission media refers to the physical means—cables of various types, radio tower, satellite, or ether (the medium that transmit radio waves)—that allow the transmission of telecommunication signals. The question to ask here is: *Which medium is used to transmit data?*

2.2.6 Information Exchange Media

Information exchange media include all data media used to transport information, e.g., all storage and transmission media. The question to ask here is: *Which data medium is used to exchange information between different locations?*

For example, information can be exchanged by storing it on a removable medium and transporting the medium from one location to another. These storage media include microfilms, paper, and floppy disks. Information can also be exchanged directly, if transmission media such as coaxial cables, optical fibers, or radio waves are used.

2.2.7 Presentation Spaces and Presentation Values

The terms described above serve as a basis to characterize the term medium in the information processing context. The description of perception media is closest to our definition of media: those media concerned mainly with the human senses. Each medium defines presentation values in presentation spaces [HD90, SH91], which address our five senses.

Paper or computer monitors are examples of visual presentation spaces. A computer-controlled slide show that projects a screen's content over the entire projection screen is a visual presentation space. Stereophony and quadrophony define acoustic presentation spaces. Presentation spaces are part of the above-described presentation media used to output information.

Presentation values determine how information from various media is represented. While text is a medium that represents a sentence visually as a sequence of characters, voice is a medium that represents information acoustically in the form of pressure waves. In some media, the presentation values cannot be interpreted correctly by humans. Examples include temperature, taste, and smell. Other media require a predefined set of symbols we have to learn to be able to understand this information. This class includes text, voice, and gestures.

Presentation values can be available as a continuous sequence or as a sequence of single values. Fluctuations in pressure waves do not occur as single values; they define acoustic signals. Electromagnetic waves in the range perceived by the human eye are not scanned with regard to time, which means that they form a continuum. The characters of a piece of text and the sampling values of an audio signal are sequences composed of single values.

2.2.8 Presentation Dimensions

Each presentation space has one or more presentation dimensions. A computer monitor has two space dimensions, while holography and stereophony need a third one. Time can occur as an additional dimension within each presentation space, which is critical for multimedia systems. Media are classified in two categories with regard to the time dimensions of their presentation space:

1. Text, graphics, and pictures are called discrete media, as they are composed of time-independent information items. Indeed, they may be displayed according to a wide variety of timing or even sequencing, and still remain meaningful. We say that time is not part of the semantics of discrete media. The term discrete tends to blur, as modern computer-based text and graphics presentations are often value-discrete and time-continuous. For example, the text of a book is a discrete medium. Each method used to process discrete media should be as fast as possible. On the other hand, time is not the critical factor, because the validity (and thus the correctness) of data does not depend on a time condition (at least not within a time frame of seconds or less). We could also speak about longer of shorter time conditions.

2. Continuous media refers to sound or motion video, where the presentation requires a continuous playout as time passes. In other words, time, or more exactly time-dependency between information items, is part of the information itself. If the timing is changed, or the sequencing of the items modified, the meaning is altered. We say that time is part of the semantics of continuous media. Continuous media are also called time-dependent media. Another technical consequence when dealing with continuous media is that they also require the networks that carry them to respect this time-dependency.

How these media are processed is time-critical, because the validity (correctness) of data depends on a time condition. If an audio sampling value is transmitted too late it may become invalid or wrong, for example, since the audio data that follow this value have already been played out over the loudspeaker. In audio and video, the representation values form a continuous sequence, where video means pure moving images. A combination of audio and moving images, like in television or movies, is not synonymous with the term video. For this reason, they are called continuous media. When time-dependent representation values that occur aperiodically are distinguished, they are often not put under the continuous media category. For a multimedia system, we also have to consider such non-continuous sequences of representation values. This type of representation-value sequence occurs when information is captured by use of a pointer (e.g., a mouse) and transmitted within cooperative applications using a common screen window. Here, the continuous medium and time-dependent medium are synonymous. By this definition, continuous media are video (moving images) of natural or artificial origin, audio, which is normally stored as a sequence of digitized pressure-wave samples, and signals from various sensors, such as air pressure, temperature, humidity, pressure, or radioactivity sensors.

The terms that describe a temporally discrete or continuous medium do not refer to the internal data representation, for example, in the way the term representation medium has been introduced. They refer to the impression that the viewer or auditor gets. The example of a movie shows that continuous-media data often consist of a sequence of discrete values, which follow one another within the representation space as a function of time. In this example, a sequence of at least 16 single images per second gives the impression of continuity, which is due to the perceptual mechanisms of the human eye.

Based on word components, we could call any system a multimedia system that supports more than one medium. However, this characterization falls short as it provides only a quantitative evaluation. Each system could be classified as a multimedia system that processes both text and graphics media. Such systems have been available for quite some time, so that they would not justify the newly coined term. The term multimedia is more of a qualitative than a quantitative nature.

As defined in [SRR90, SH91], the number of supported media is less decisive than the type of supported media for a multimedia system to live up to its name. Note that there is controversy about this definition. Even standardization bodies normally use a coarser interpretation.

2.3 Key Properties of a Multimedia System

Multimedia systems involve several fundamental notions. They must be computer-controlled. Thus, a computer must be involved at least in the presentation of the information to the user. They are integrated, that is, they use a minimal number of different devices. An example is the use of a single computer screen to display all types of visual information. They must support media independence. And lastly, they need to handle discrete and continuous media. The following sections describe these key properties.

2.3.1 Discrete and Continuous Media

Not just any arbitrary combination of media deserves the name multimedia. Many people call a simple word processor that handles embedded graphics a multimedia application because it uses two media. By our definition, we talk about multimedia if the application uses both discrete and continuous media. This means that a multimedia application should process at least one discrete and one continuous medium. A word processor with embedded graphics is not a multimedia application by our definition.

2.3.2 Independent Media

An important aspect is that the media used in a multimedia system should be independent. Although a computer-controlled video recorder handles audio and moving image information, there is a temporal dependence between the audio part and the video part. In contrast, a system that combines signals recorded on a DAT (Digital Audio Tape) recorder with some text stored in a computer to create a presentation meets the independence criterion. Other examples are combined text and graphics blocks, which can be in an arbitrary space arrangement in relation to one another.

2.3.3 Computer-Controlled Systems

The independence of media creates a way to combine media in an arbitrary form for presentation. For this purpose, the computer is the ideal tool. That is, we need a system capable of processing media in a computer-controlled way. The system can be optionally programmed by a system programmer and/or by a user (within certain limits). The simple recording or playout of various media in a system, such as a video recorder, is not sufficient to meet the computer-control criterion.

2.3.4 Integration

Computer-controlled independent media streams can be integrated to form a global system so that, together, they provide a certain function. To this end, synchronic relationships of time, space, and content are created between them. A word processor that supports text, spreadsheets, and graphics does not meet the integration criterion unless it allows program-supported references between the data. We achieve a high degree of integration only if the application is capable of, for example, updating graphics and text elements automatically as soon as the contents of the related spreadsheet cell changes.

This kind of flexible media handling is not a matter to be taken for granted—even in many products sold under the multimedia system label. This aspect is important when talking of integrated multimedia systems. Simply speaking, such systems should allow us to do with moving images and sound what we can do with text and graphics [AGH90]. While conventional systems can send a text message to another user, a highly integrated multimedia system provides this function *and* support for voice messages or a voice-text combination.

2.3.5 Summary

Several properties that help define the term multimedia have been described, where the media are of central significance. This book describes networked multimedia systems. This is important as almost all modern computers are connected to communication networks. If we study multimedia functions from a local computer's perspective, we take a step backwards. Also, distributed environments offer the most interesting multimedia applications as they enable us not only to create, process, represent, and store multimedia information, but to exchange them beyond the limits of our computers.

Finally, continuous media require a changing set of data in terms of time, that is, a data stream. The following section discusses data streams.

2.4 Characterizing Data Streams

Distributed networked multimedia systems transmit both discrete and continuous media streams, i.e., they exchange information. In a digital system, information is split into units (packets) before it is transmitted. These packets are sent by one system component (the source) and received by another one (the sink). Source and sink can reside on different computers. A data stream consists of a (temporal) sequence of packets. This means that it has a time component and a lifetime.

Packets can carry information from continuous and discrete media. The transmission of voice in a telephone system is an example of a continuous medium. When we transmit a text file, we create a data stream that represents a discrete medium.

When we transmit information originating from various media, we obtain data streams that have very different characteristics. The attributes asynchronous, synchronous, and isochronous are traditionally used in the field of telecommunications to describe the characteristics of a data transmission. For example, they are used in FDDI to describe the set of options available for an end-to-end delay in the transmission of single packets.

2.4.1 Asynchronous Transmission Mode

In the broadest sense of the term, a communication is called asynchronous if a sender and receiver do not need to coordinate before data can be transmitted. In asynchronous transmission, the transmission may start at any given instant. The bit synchronization that determines the start of each bit is provided by two independent clocks, one at the sender, the other at the receiver. An example of asynchronous transmission is the way in which simple ASCII terminals are usually attached to host computers. Each time the character "A" is pressed, a sequence of bits is generated at a preset speed. To inform the computer interface that a character is arriving, a special signal called the start signal—which is not necessarily a bit—precedes the information bits. Likewise, another special signal, the stop signal, follows the last information bit.

2.4.2 Synchronous Transmission Mode

The term synchronous refers to the relationship of two or more repetitive signals that have simultaneous occurrences of significant instants. In synchronous transmission, the beginning of the transmission may only take place at well-defined times, matching a clocking signal that runs the synchronism with that of the receiver. To see why clocked transmission is important, consider what might happen to a digitized voice signal when it is transferred across a nonsynchronized network. As more traffic enters the network, the transmission of a given signal may experience increased delay. Thus, a data stream moving across a network might slow down temporarily when other traffic enters the network and then speed up again when the traffic subsides. If audio from a digitized phone call is delayed, however, the human listening to the call will hear the delay as annoying interference or noise. Once a receiver starts to play digitized samples that arrive late, the receiver cannot speed up the playback to catch up with the rest of the stream.

2.4.3 Isochronous Transmission Mode

The term isochronous refers to a periodic signal, pertaining to transmission in which the time interval separating any two corresponding transitions is equal to the unit interval or to a multiple of the unit interval. Secondly, it refers to data transmission in which corresponding significant instants of two or more sequential signals have a constant phase relationship. This mode is a form of data transmission in which individual characters are only separated by a whole number of bit-length intervals, in contrast to asynchronous transmission, in which the characters may be separated by random-length intervals. For example, an end-to-end network connection is said to be isochronous if the bit rate over the connection is guaranteed and if the value of the delay jitter is also guaranteed and small. The notion of isochronism serves to describe what the performance of a network should be in order to satisfactorily transport continuous media streams, such as real-time audio or motion video. What is required to transport audio and video in real time? If the source transmits bits at a certain rate, the network should be able to meet that rate in a sustained way.

These three attributes are a simplified classification of different types of data streams. The following sections describe other key properties.

2.5 Characterizing Continuous Media Data Streams

This section provides a summary of the characteristics for data streams that occur in multimedia systems in relation to audio and video transmissions. The description includes effects of compression methods applied to the data streams before they are transmitted. This classification applies to distributed and local environments.

2.5.1 Strongly and Weakly Periodic Data Streams

The first property of data streams relates to the time intervals between fully completed transmissions of consecutive information units or packets. Based on the moment in which the packets become ready, we distinguish between the following variants:

- When the time interval between neighboring packets is constant, then this data stream is called a strongly periodic data stream. This also means that there is minimal jitter—ideally zero. Figure 2-1 shows such a data stream. An example for this type is PCM-encoded (Pulse Code Modulation) voice in telephone systems.

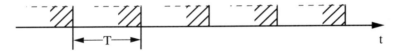

Figure 2-1 Strongly periodic data stream; time intervals have the same duration between consecutive packets.

- The duration of the time intervals between neighboring packets is often described as a function with finite period duration. However, this time interval is not constant between neighboring packets (or it would be a strongly periodic data stream). In the case shown in Figure 2-2, we speak of a weakly periodic data stream.

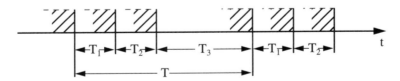

Figure 2-2 Weakly periodic data stream; time intervals between consecutive packets are periodic.

- All other transmission options are called aperiodic data streams, which relates to the sequence of time interval duration, as shown in Figure 2-3.

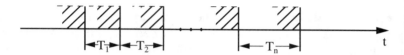

Figure 2-3 Aperiodic data stream; the time interval sequence is neither constant nor weakly periodic.

An example of an aperiodic data stream is a multimedia conference application with a common screen window. Often, the status (left button pressed) and the current coordinates of the mouse moved by another user have to be transmitted to other participants. If this information were transmitted periodically, it would cause a high data rate and an extremely high redundancy. The ideal system should transmit only data within the active session that reflect a change in either position or status.

2.5.2 Variation of the Data Volume of Consecutive Information Units

A second characteristic to qualify data streams concerns how the data quantity of consecutive information units or packets varies.

- If the quantity of data remains constant during the entire lifetime of a data stream, then we speak of a strongly regular data stream. Figure 2-4 shows such a data stream. This characteristic is typical for an uncompressed digital audio-video stream. Practical examples are a full-image encoded data stream delivered by camera or an audio sequence originating from an audio CD.

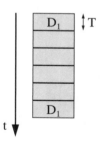

Figure 2-4 Strongly regular data stream; the data quantity is constant in all packets.

- If the quantity of data varies periodically (over time), then this is a weakly regular data stream. Figure 2-5 shows an example.

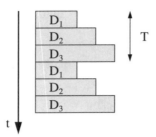

Figure 2-5 Weakly regular data stream; the packets' data stream varies periodically.

Some video compression methods use a concept that encodes and compresses full images individually. The resulting information unit is a relatively large data packet in the data stream. For reasons of simplicity, we will not consider the packet length, which is limited during the transmission, depending on the communication layer. These packets are transmitted periodically, e.g., every two seconds. For all images between two single images of the video stream, the differences between two consecutive images each form the information that is actually transmitted.

An example is MPEG (see Section 7.7), where I images are compressed single images, while the compression of P images and B images uses only image differences, so that the data volume is much smaller. No constant bit rate is defined for compressed I, P, and B packets. However, a typical average $I:B:P$ ratio of the resulting data quantities is 10:1:2, which results in a weakly regular data stream over the long-term average.

Data streams are called irregular when the data quantity is neither constant, nor changing by a periodic function (see Figure 2-6). This data stream is more difficult to transmit and process compared to the variants described earlier.

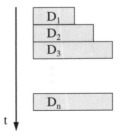

Figure 2-6 Irregular data stream; the packets' data quantity is not constant and does not vary periodically.

When applying a compression method that creates a data stream with a variable bit rate, the size of the single information units (each derived from a single image) is determined from the image content that has changed in respect to the previous image. The size of the resulting information units normally depends on the video sequence and the data stream is irregular.

2.5.3 Interrelationship of Consecutive Packets

The third qualification characteristic concerns the continuity or the relationship between consecutive packets. Are packets transmitted progressively, or is there a gap between packets? We can describe this characteristic by looking at how the corresponding resource is utilized. One such resource is the network.

- Figure 2-7 shows an interrelated information transfer. All packets are transmitted one after the other without gaps in between. Additional or layer-independent information to identify user data is included, e.g., error detection codes. This means that a specific resource is utilized at 100 percent. An interrelated data stream allows maximum throughput and achieves optimum utilization of a resource. An ISDN B channel that transmits audio data at 64 Kbit/s is an example.

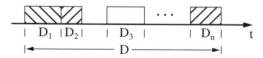

Figure 2-7 Interrelated data stream; packets are transmitted without gaps in between.

- The transmission of an interrelated data stream over a higher-capacity channel causes gaps between packets. Each data stream that includes gaps between its information units is called a non-interrelated data stream. Figure 2-8 shows an example. In this case, it is not important whether or not there are gaps between all packets or whether the duration of the gaps varies. An example of a non-interrelated data stream is the transmission of a data stream encoded by the DVI-PLV method over an FDDI network. An average bit rate of 1.2 Mbit/s leads inherently to gaps between some packets in transit.

Figure 2-8 Non-interrelated data stream; there are gaps between packets.

To better understand the characteristics described above, consider the following example:

A PAL video signal is sampled by a camera and digitized in a computer. No compression is applied. The resulting data stream is strongly periodic, strongly regular, and interrelated, as shown in Figure 2-4. There are no gaps between packets. If we use the MPEG method for compression, combined with the digitizing process, we obtain a weakly periodic and weakly regular data stream (referring to its longer duration). And if we assume we use a 16-Mbit/s token-ring network for transmission, our data stream will also be noninterrelated.

2.6 Information Units

Continuous (time-dependent) media consist of a (temporal) sequence of information units. Based on Protocol Data Units (PDUs), this section describes such an information unit, called a Logical Data Unit (LDU). An LDU's information quantity and data quantities can have different meanings:

1. Let's use Joseph Haydn's symphony, *The Bear*, as our first example. It consists of the four musical movements, *vivace assai*, *allegretto*, *menuet*, and *finale vivace*. Each movement is an independent, self-sufficient part of this composition. It contains a sequence of scores for the musical instruments used. In a digital system, these scores are a sequence of sampling values. We will not use any compression in this example, but apply PCM encoding with a linear characteristic curve. For CD-DA quality, this means 44,100 sampling values per second, which are encoded at 16 bits per channel. On a CD, these sampling values are grouped into units with a duration of 1/75 second. We could now look at the entire composition and define single movements, single scores, the grouped 1/75-s sampling values, or even single sampling values as LDUs. Some operations can be applied to the playback of the entire composition—as one single LDU. Other functions refer to the smallest meaningful unit (in this case the scores). In digital signal processing, sampling values are LDUs.

2. In Figure 2-9, we see that the uncompressed video sequence consists of single chips, each representing a scene. Each of these scenes consists of a sequence of single images. Each single image can be separated into various regions, for example regions with a size of 16×16 pixels. In turn, each pixel contains a luminance value and a chrominance value.

   This means that a single image is not the only possible LDU in a motion video sequence. Each scene and each pixel are also LDUs. The redundancies in single image sequences of an MPEG-encoded video stream can be used to reduce the data quantity by applying an inter-frame compression method. In this case, the smallest self-sufficient meaningful units are single-image sequences.

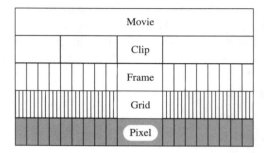

Figure 2-9 Granularity of a motion video sequence showing its logical data units (LDUs).

A phenomenon called granularity characterizes the hierarchical decomposition of an audio or video stream in its components. This example uses a symphony and a motion video to generally describe extensive information units. We distinguish between closed and open LDUs. Closed LDUs have a well-defined duration. They are normally stored sequences. In open LDUs, the data stream's duration is not known in advance. Such a data stream is delivered to the computer by a camera, a microphone, or a similar device.

The following chapter builds on these fundamental characteristics of a multimedia system to describe audio data in more detail, primarily concentrating on voice processing.

Chapter 3

Audio Technology

Audiology is the discipline interested in manipulating acoustic signals that can be perceived by humans. Important aspects are psychoacoustics, music, the MIDI (Musical Instrument Digital Interface) standard, and speech synthesis and analysis. Most multimedia applications use audio in the form of music and/or speech, and voice communication is of particular significance in distributed multimedia applications.

In addition to providing an introduction to basic audio signal technologies and the MIDI standard, this chapter explains various enabling schemes, including speech synthesis, speech recognition, and speech transmission [Loy85, Fla72, FS92, Beg94, OS90', Fal85, Bri86, Ace93, Sch92]. In particular, it covers the use of sound, music, and speech in multimedia, for example, formats used in audio technology, and how audio material is represented in computers [Boo87, Tec89].

Chapter 8 covers storage of audio data (and other media data) on optical disks because this technology is not limited to audio signals. The compression methods used for audio and video signals are described in Chapter 9 because many methods available for different media to encode information are similar.

3.1 What Is Sound?

Sound is a physical phenomenon caused by vibration of material, such as a violin string or a wood log. This type of vibration triggers pressure wave fluctuations in the air around the material. The pressure waves propagate in the air. The pattern of this oscillation (see Figure 3-1) is called wave form [Tec89]. We hear a sound when such a wave reaches our ears.

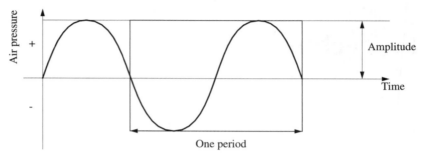

Figure 3-1 Pressure wave oscillation in the air.

This wave form occurs repeatedly at regular intervals or periods. Sound waves have a natural origin, so they are never absolutely uniform or periodic. A sound that has a recognizable periodicity is referred to as music rather than sound, which does not have this behavior. Examples of periodic sounds are sounds generated by musical instruments, vocal sounds, wind sounds, or a bird's twitter. Non-periodic sounds are, for example, drums, coughing, sneezing, or the brawl or murmur of water.

3.1.1 Frequency

A sound's *frequency* is the reciprocal value of its period. Similarly, the frequency represents the number of periods per second and is measured in hertz (Hz) or cycles per second (cps). A common abbreviation is kilohertz (kHz), which describes 1,000 oscillations per second, corresponding to 1,000 Hz [Boo87].

Sound processes that occur in liquids, gases, and solids are classified by frequency range:

- Infrasonic: 0 to 20 Hz
- Audiosonic: 20 Hz to 20 kHz
- Ultrasonic: 20 kHz to 1 GHz
- Hypersonic: 1 GHz to 10 THz

Sound in the audiosonic frequency range is primarily important for multimedia systems. In this text, we use audio as a representative medium for all acoustic signals in this frequency range. The waves in the audiosonic frequency range are also called acoustic signals [Boo87]. Speech is the signal humans generate by use of their speech organs. These signals can be reproduced by machines. For example, music signals have frequencies in the 20 Hz to 20 kHz range. We could add noise to speech and music as another type of audio signal. Noise is defined as a sound event without functional purpose, but this is not a dogmatic definition. For instance, we could add unintelligible language to our definition of noise.

3.1.2 Amplitude

A sound has a property called amplitude, which humans perceive subjectively as loudness or volume. The amplitude of a sound is a measuring unit used to deviate the pressure wave from its mean value (idle state).

3.1.3 Sound Perception and Psychoacoustics

The way humans perceive sound can be summarized as a sequence of events: Sound enters the ear canal. At the eardrum, sound energy (air pressure changes) are transformed into mechanical energy of eardrum movement. The outer ear comprises the pinna, which is composed of cartilage and has a relatively poor blood supply. Its presence on both sides of the head allows us to localize the source of sound from the front versus the back. Our ability to localize from side to side depends on the relative intensity and relative phase of sound reaching each ear and the analysis of the phase/intensity differences within the brainstem. The cochlea is a snail-shaped structure that is the sensory organ of hearing. The vibrational patterns that are initiated by vibration set up a traveling wave pattern within the cochlea. This wavelike pattern causes a shearing of the cilia of the outer and inner hair cells. This shearing causes hair cell depolarization resulting in on/off neural impulses that the brain interprets as sound.

Psychoacoustics is a discipline that studies the relationship between acoustic waves at the auditory ossicle and the spatial recognition of the auditor. We distinguish between two main perspectives, described briefly in the following sections.

3.1.3.1 The Physical Acoustic Perspective

Figure 3-2 is a simplified schematic representation of an auditor who perceives sound. Sound from a sound source diffuses in concentric pressures waves. The position of the source can be described by the distance to the center of the auditor's head and by two angles: one in the horizontal and one on the vertical level. It is obvious that sound waves originating from a source arranged on the right side of the auditor reaches the right ear earlier than the left one. The time difference of sound waves reaching the ears is called interaural time difference (ITD). If a sound source prevails almost totally on one side, i.e., at an angle of 90 degrees from the auditor, then ITD reaches a maximum of approximately 0.7ms to 0.8ms. If the sound source is close, i.e., if the distance between the source and the auditor is very short, then the interaural intensity difference (IID) differs significantly from zero. These two characteristics of sound waves arriving at the ear can be measured and described and represent the basis for spatial recognition [Ken95].

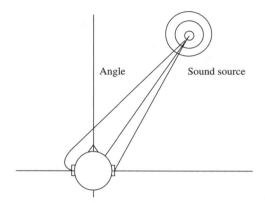

Figure 3-2 The basics of sound perception.

An important property of the basic hearing process can be determined in the frequency range. When sound waves reach the auditor's ear, they change due to the interaction between the original wave and the auditor's body. For example, in the 3-kHz range, there is a strong resonance in the perception of size caused by the resonance in the hearing canal. These properties can be measured and documented as a head-related transfer function (HRTF) [Bla74, GM94]. Since the interaction between the sound waves and the auditor's body is complex, it generates a strong dependence on the horizontal and vertical angles, in which the sound source is arranged.

Another important aspect in spatial sound recognition is the physical acoustics of natural sounds. Similarly to all systems based on the wave model, the laws of reflection, refraction, and dispersion apply to the longitudinal diffusion of sound waves in air (these waves reach a speed of 344m/s). In a closed room, each surface reflects sound waves. All waves will eventually reach the human ear, reflected many times on the way from the sound source to the ear. Sound that reaches the ear directly includes information about the horizontal and vertical angles of the sound source. Other waves, derived from direct waves, arrive later and provide additional information about the room's properties. The relationship between direct and derived waves can be used to get an idea about the distance between source and auditor. This allows us to expand and apply basic localization rules to moving sound sources or auditors. To achieve this expansion, we measure the sound speed, which is also called the doubling effect. The sound spectrum of a sound source moving towards an auditor drifts upwards, while the source spectrum moving away from an auditor drifts downwards. It is not difficult to determine this effect; an everyday example is when we are passed by a piercing ambulance siren.

3.1.3.2 The Psychoacoustic Perspective

One of the fundamental properties of humans' spatial hearing perception is the so-called first wave-front law. This law says that an auditor's judgment about the direction of an acoustic event is primarily influenced by the sound that takes the shortest and most direct way. In a test environment [Bla71], a listener was seated in front of two loudspeakers of identical make. Although both issued sound at the same amplitude, the listener localized it stronger on the right side, because the left loudspeaker transmitted with a delay of approximately 15 ms. When the delay exceeded 50 ms, the listener perceived two different sound events from the left and right loudspeakers. To compensate for the effect produced by this delay, we can increase the amplitude of the delayed channel (Haas effect).

As with all human perception channels, the ear's cochlea transforms stimulation logarithmically. The size of a sound pressure level (SPL) is measured in decibels. An audibility threshold value of 20 microPascal is the limit value above which a sound can just about be perceived. This value functions as a basis of the sound pressure, measured in decibels. The dynamic range of the ear's sound recognition is in the range of up to 130 dB.

Sound example	Sound pressure size
Rustling of paper	20 dB
Spoken language	60 dB
Heavy road traffic	80 dB
Rock band	120 dB
Pain sensitivity threshold	130 dB

Table 3-1 Various sound pressure examples.

The perception sensitivity we call loudness is not linear across all frequencies and intensities. The Fletcher-Munson graphics (of 1933) show the sound intensity required to enable the perception of constant loudness. Human sound perception is most sensitive in the mean frequency ranges between 700 Hz and approximately 6,600 Hz. The equal loudness curves of Fletcher and Munson (see Figure 3-3) show that the human hearing system responds much better to the mean frequency range than it does to low and very high frequencies.

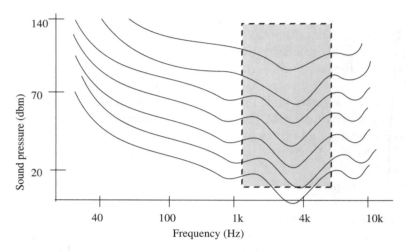

Figure 3-3 Equal loudness curves (Fletcher and Munson).

Human aural perception uses a subjective tone pitch, which is not proportional to the frequency. With dark tones, two tones that differ by a factor of two in the frequency scale correspond to exactly one octave. In higher frequencies, such a frequency mixture is sensed as a smaller interval.

In special situations, some parts of an acoustic event can be measured, although they cannot be heard. The reason is that part of a sound mixture masks another part. This masking effect can be observed in the time and frequency ranges. It is important to understand this effect, because it contributes to encoding and compression techniques (see Chapter 7) applied to sound signals. For example, an auditor will perceive two different waves when a loud and a lower sinus wave have very different frequencies. On the other hand, if the lower wave is near the frequency of the louder wave, then the lower will no longer be heard, that is, it will fall below the frequency masking threshold. In this case, the auditor will no longer perceive the lower wave. On the other hand, a loud gun shot will mask lower sounds in the time range several seconds after they occurred.

3.2 Audio Representation on Computers

Before the continuous curve of a sound wave can be represented on a computer, the computer has to measure the wave's amplitude in regular time intervals. It then takes the result and generates a sequence of sampling values, or samples for short. Figure 3-4 shows the period of a digitally sampled wave.

Audio Representation on Computers

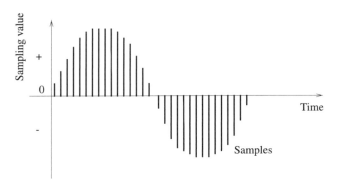

Figure 3-4 Sampling a wave.

The mechanism that converts an audio signal into a sequence of digital samples is called an analog-to-digital converter (ADC) and a digital-to-analog converter (DAC) is used to achieve the opposite conversion.

3.2.1 Sampling Rate

The rate at which a continuous wave form is sampled (see Figure 3-4) is called the sampling rate. Like frequency, the sampling rate is measured in Hz. For example, CDs are sampled at a rate of 44,100 Hz, which may appear to be above the frequency range perceived by humans. However, the bandwidth—in this case, 20,000 Hz − 20 Hz = 19,980 Hz—that can represent a digitally sampled audio signal is only about half as big as a CD's sampling rate, because CDs use the Nyquist sampling theorem. This means that a sampling rate of 44,100 Hz covers only frequencies in the range from 0 Hz to 22,050 Hz. This limit is very close to the human hearing capability.

3.2.2 Quantization

The digitization process requires two steps. First the analog signal must be sampled. This means that only a discrete set of values is retained at (generally regular) time or space intervals. The second step involves quantization. The quantization process consists of converting a sampled signal into a signal that can take only a limited number of values. An 8-bit quantization provides 256 possible values, while a 16-bit quantization in CD quality results in more than 65,536 possible values. Figure 3-5 shows a 3-bit quantization.

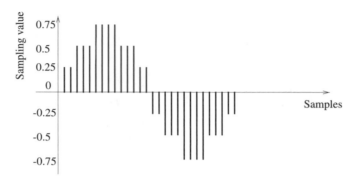

Figure 3-5 3-bit quantization.

The values transformed by a 3-bit quantization process can accept eight different characteristics: 0.75, 0.5, 0.25, 0, -0.25, -0.5, -0.75, and -1, so that we obtain an "angular-shape" wave. This means that the lower the quantization (in bits), the more the resulting sound quality deteriorates.

3.3 Three-Dimensional Sound Projection

The invention of loudspeakers in the 1920s had roughly the same effect on audio processing as the first light bulb did on everyday life. Suddenly, it was possible to play sounds in any kind of room. The first sound playback before an auditorium can be compared to the first movies. After some experimentation time with various sets of loudspeakers, it was found that the use of a two-channel sound (stereo) produces the best hearing effect for most people. Among the large number of home applications that have been developed over the years—for example, radio or music records—there were also movie systems and modern music components using a larger number of loudspeakers [Cho71] to better control spatial sound effects. Current multimedia and virtual-reality implementations that include such components show a considerable concentration on spatial sound effects and three-dimensional sound projections [Beg94].

3.3.1 Spatial Sound

Figure 3-6 shows schematically how an auditor perceives three-dimensional spatial sound in a closed room. As an expansion of Figure 3-2, the walls and other objects in a room absorb and reflect the sound's dispersion paths.

Three-Dimensional Sound Projection 29

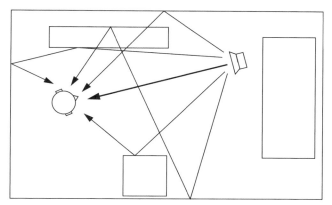

Figure 3-6 Sound dispersion in a closed room.

The shortest path between the sound source and the auditor is called the direct sound path (bold line in Figure 3-6). This path carries the first sound waves towards the auditor's head. All other sound paths are reflected, which means that they are temporally delayed before they arrive at the auditor's ear. These delays are related to the geometric length of the sound paths inside the room and they usually occur because the reflecting sound path is longer than the direct one [Ken95].

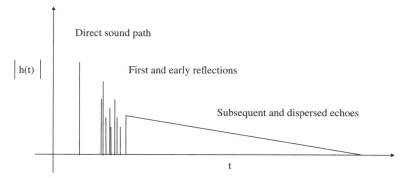

Figure 3-7 Pulse response in a closed room.

Figure 3-7 shows the energy of a sound wave arriving at an auditor's ear as a function over time. The source's sound stimulus has a pulse form (e.g., like a gun shot), so that the energy of the direct sound path appears as a large swing in the figure. This swing is called the pulse response. All subsequent parts of the pulse response refer to reflecting sound paths. Very early reflections belong to the group of paths that are reflected only once. This sound path group can occur in almost all natural environments where echoes occur. Subsequent and dispersed echoes represent a bundle of sound paths that have been reflected several times. These paths cannot be isolated in the pulse response because the pulse is diffuse and diverted both in the time and in the frequency range. All sound paths leading to the human ear are additionally influenced by the auditor's individual HRTF (head-related transfer function). HRTF is a function of the path's direction (horizontal and vertical angles) to the auditor (see also Figure 3-2). A sound projection that considers HRTF for each sound path, for example by use of headphones, is also called binaural stereophony [KDS95].

3.3.2 Reflection Systems

Spatial sound systems are used in many different applications, and each of them has different requirements. A rough classification (see Table 3-2) groups these applications into a scientific and a consumer-oriented approach. In general, the scientific approach (simulation) is more common than consumer-oriented (imitation).

Approach	Attribute	Applications
scientific	simulation, precise, complex, offline	research, architecture, computer music
consumer-oriented	imitation, unprecise, impressive, real-time	cinema, music, home movies, computer games

Table 3-2 Reflection system applications.

The scientific approach uses simulation options that enable professionals such as architects to predict the acoustics of a room when they plan a building on the basis of CAD models [Vor89, KDS95]. The calculations required to generate a pulse response of a hearing situation based on the database of a CAD model are complex, and normally only experts hear the results output by such systems. Consumer systems concentrate mainly on applications that create a spatial or a virtual environment. In cinemas, for example, a special multi-channel sound technique is used to create special sound effects. In modern pop music, special echo processes based on signal feedback algorithms are used [Moo79] to perfect the sound of recorded clips. Modern multimedia environments use both approaches [JLW95]. Special interactive environments for images and computer games, including sound or modern art projects, can be implemented, but in order to create a spatial sound, they require state-of-the-art technologies.

3.4 Music and the MIDI Standard

We know from previous sections that any sound can be represented as a digitized sound signal that is a sequence of samples, each encoded with binary digits. This sequence may be uncompressed as on audio compact disks, or compressed. We also know that any sound may be represented in that way, including music. A characteristic of this representation mode is that it does not preserve the sound's semantic description. Unless complex recognition techniques are used, the computer does not know whether a bit sequence represents speech or music, for example, and if music what notes are used and by which instrument.

Music can be described in a symbolic way. On paper, we have the full scores. Computers and electronic musical instruments use a similar technique, and most of them employ the Musical Instrument Digital Interface (MIDI), a standard developed in the early 1980s. The MIDI standard defines how to code all the elements of musical scores, such as sequences of notes, timing conditions, and the instrument to play each note.

3.4.1 Introduction to MIDI

MIDI represents a set of specifications used in instrument development so that instruments from different manufacturers can easily exchange musical information [Loy85]. The MIDI protocol is an entire music description language in binary form. Each word describing an action of a musical performance is assigned a specific binary code. A MIDI interface is composed of two different components:

- Hardware to connect the equipment. MIDI hardware specifies the physical connection of musical instruments. It adds a MIDI port to an instrument, it specifies a MIDI cable (that connects two instruments), and processes electrical signals received over the cable.
- A data format that encodes information to be processed by the hardware. The MIDI data format does not include the encoding of individual sampling values, such as audio data formats. Instead, MIDI uses a specific data format for each instrument, describing things like the start and end of scores, the basis frequency, and loudness, in addition to the instrument itself.

The MIDI data format is digital and data are grouped into MIDI messages. When a musician plays a key, the MIDI interface generates a MIDI message that defines the start of each score and its intensity. This message is transmitted to machines connected to the system. As soon as the musician releases the key, another signal (MIDI message) is created and transmitted.

3.4.2 MIDI Devices

An instrument that complies with both components defined by the MIDI standard is a MIDI device (e.g., a synthesizer) able to communicate with other MIDI devices over channels. The MIDI standard specifies 16 channels. A MIDI device is mapped onto a channel. Musical data transmitted over a channel are reproduced in the synthesizer at the receiver's end. The MIDI standard identifies 128 instruments by means of numbers, including noise effects (e.g., a phone ringing or an airplane take-off). For example 0 specifies a piano, 12 a marimba, 40 a violin, and 73 a flute.

Some instruments enable a user to play one single score (e.g., a flute) exclusively, while other instruments allow concurrent playing of scores (e.g., an organ). The maximum number of scores that can be played concurrently is an important property of synthesizers. This number can vary between 3 and 16 scores per channel.

A computer uses the MIDI interface to control instruments for playout. The computer can use the same interface to receive, store, and process encoded musical data. In the MIDI environment, these data are generated on a keyboard and played out by a synthesizer—the heart of each MIDI system. A typical synthesizer is similar to a regular piano keyboard, but it has an additional operating element (for detailed information see [Boo87]). A sequencer is used to buffer or modify these data. In a multimedia application, the sequencer resides in the computer.

3.4.3 The MIDI and SMPTE Timing Standards

The MIDI clock is used by a receiver to synchronize itself to the sender's clock. To allow synchronization, 24 identifiers for each quarter note are transmitted. Alternatively, the SMPTE (Society of Motion Picture and Television Engineers) timing code can be sent to allow receiver-sender synchronization. SMPTE defines a frame format by hours:minutes:seconds:, for example 30 frames/s. This information is transmitted in a rate that would exceed the bandwidth of existing MIDI connections. For this reason, the MIDI time code is normally used for synchronization because it does not transmit the entire time representation of each frame.

3.5 Speech Signals

Speech can be processed by humans or machines, although it is the dominant form of communication of human beings. The field of study of the handling of digitized speech is called digital speech processing.

3.5.1 Human Speech

Speech is based on spoken languages, which means that it has a semantic content. Human beings use their speech organs without the need to knowingly control the generation of sounds. (Other species such as bats also use acoustic signals to transmit information, but we will not discuss this here.) Speech understanding means the efficient adaptation to speakers and their speaking habits. Despite the large number of different dialects and emotional pronunciations, we can understand each other's language. The brain is capable of achieving a very good separation between speech and interference, using the signals received by both ears. It is much more difficult for humans to filter signals received in one ear only. The brain corrects speech recognition errors because it understands the content, the grammar rules, and the phonetic and lexical word forms.

Speech signals have two important characteristics that can be used by speech processing applications:

- Voiced speech signals (in contrast to unvoiced sounds) have an almost periodic structure over a certain time interval, so that these signals remain quasi-stationary for about 30 ms.
- The spectrum of some sounds have characteristic maxima that normally involve up to five frequencies. These frequency maxima, generated when speaking, are called formants. By definition, a formant is a characteristic component of the quality of an utterance.

[All85, BN93] describe and model human speech.

3.5.2 Speech Synthesis

Computers can translate an encoded description of a message into speech. This scheme is called speech synthesis. A particular type of synthesis is text-to-speech conversion. Fair-quality text-to-speech software has been commercially available for various computers and workstations, although the speech produced in some lacks naturalness.

Speech recognition is normally achieved by drawing various comparisons. With the current technology, a speaker-dependent recognition of approximately 25,000 words is possible. The problems in speech recognition affecting the recognition quality include dialects, emotional pronunciations, and environmental noise. It will probably take some time before the considerable performance discrepancy between the human brain and a powerful computer will be bridged in order to improve speech recognition and speech generation [Ace93, Mam93].

3.6 Speech Output

Speech output deals with the machine generation of speech. Considerable work [End84, Fel85] has been achieved in this field. As early as in the middle of the 19th century, Helmholtz used several connected mechanical resonators to simulate the human vocal tract and generate utterances. In 1940, Dudley introduced the first "speech synthesizer," which simulated mechanical speech from electrical oscillating circles [Fal85].

A major challenge in speech output is how to generate these signals in real time for a speech output system to be able, for instance, to convert text to speech automatically. Some applications (e.g., time announcements) handle this task with a limited vocabulary, but most use an extensive if not unlimited vocabulary.

The speech a machine outputs has to be understandable and should sound natural. In fact, understandability is compulsory and naturalness a nice thing to have to increase user acceptance.

It is important to understand the most important technical terms used in relation to speech output, including:

- Speech basic frequency means the lowest periodic signal share in the speech signal. It occurs in voiced sounds.
- A phoneme is a member of the set of the smallest units of speech that serve to distinguish one utterance from another in a language or dialect. It is the smallest meaningful linguistic unit but does not carry content.
- Allophones specify variants of a phoneme as a function of its phonetic environment.
- A morpheme is a meaningful linguistic unit whether in free form or bound form that contains no smaller meaningful parts. For example, house is a morpheme, while housing is not.
- A voiced sound is generated by oscillations of the vocal cords. The characters M, W, and L are examples. Voiced sounds depend strongly on the speaker.
- Unvoiced sounds are generated with the vocal cords open, for example, F and S. These sounds are relatively independent of the speaker.

3.6.1 Reproducible Speech Playout

Reproducible speech playout is a straightforward method of speech output. The speech is spoken by a human and recorded. To output the information, the stored sequence is played out. The speaker can always be recognized. The method uses a limited vocabulary or a limited set of sentences that produce an excellent output quality. The speech can be PCM-coded and stored (PCM will be described later), or one applies other data reduction methods without utilizing properties typical for speech (compression methods are described in Chapter 7).

3.6.2 Sound Concatenation in the Time Range

Speech can also be output by concatenating sounds in the time range [Ril89]. This method puts together speech units like building blocks. Such a composition can occur on various levels. In the simplest case, individual phonemes are understood as speech units. Figure 3-8 represents phonemes of the word "stairs." It is possible to generate an unlimited vocabulary with few phonemes. The problem is that there are transitions between the phonemes. Unfortunately, this problem cannot be solved entirely, but can be ameliorated with a second approach that considers allophones, or phonemes in their environment.

Speech Output

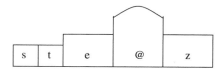

Figure 3-8 Sound concatenation of a phoneme in the time range.

If we group two phonemes we obtain a diphone. Figure 3-9 shows the word stairs again, but this time consisting of an ordered quantity of diphones.

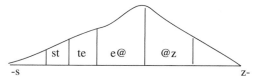

Figure 3-9 Sound concatenation of a diphone in the time range.

To further mitigate problematic transitions between speech units, we can form carrier words. We see in Figure 3-9 that speech is composed of a set of such carrier words.

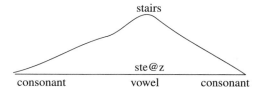

Figure 3-10 Sound concatenation of a word in the time range.

The best pronunciation of a word is achieved by storing the entire word. By previously storing a speech sequence as a whole entity, we use playout to move into the speech synthesis area (see Figure 3-10).

All these cases have a common problem that is due to transitions between speech-sound units. This effect is called coarticulation. Coarticulation means mutual sound effects across several sounds. This effect is caused by the influence of the relevant sound environment, or more specifically, by the idleness of our speech organs.

Another important factor in speech output is the so-called prosody, which describes the accentuation and melody curves of a sentence. For example, we stress words totally differently depending on whether we state something or we ask something. This means that prosody depends on the semantics of a language, so that it has to be considered in each sound concatenation in the time range [Wai88].

3.6.3 Sound Concatenation in the Frequency Range

As an alternative to concatenating sound in the time range, we can affect sound concatenation in the frequency range, for example, by formant synthesis [Ril89]. Formants are energy concentrations in the speech signal's spectrum. Formant synthesis uses filters to simulate the vocal tract. Characteristic values are the central filter frequencies and the central filter bandwidths. All voiced sounds are excited by a pulse signal with a frequency corresponding to the speech basic frequency. In contrast, unvoiced sounds are generated by a noise generator.

The characteristic values of formants define individual speech elements, such as phonemes. However, the problems incurred here are similar to those of sound concatenation in the time range. The transitions defined by a coarticulation represent a very critical aspect. Moreover, prosody has to be determined. Sound-specific methods combine a synthesis in the time and frequency ranges [Fri92b]. The results show mainly in an improved quality of both fricative and explosive sounds.

We can use a multipolar lattice filter to simulate the human vocal tract. This filter allows us to correctly model the first four or five formants. A noise generator and a radio frequency oscillator are used to simulate the vocal cords. This linear predictive coding method according to [Schr85a] is very similar to the formant synthesis.

Another approach uses tube models. The voice formation system is approximated by a simplified mechanical tube model and simulated by a digital wave filter.

3.6.4 Speech Synthesis

Speech synthesis can be used to transform an existing text into an acoustic signal [WSF92]. Figure 3-11 shows the components of such a system.

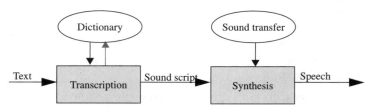

Figure 3-11 Components of a speech synthesis system, using sound concatenation in the time range.

The first step involves a transcription, or translation of the text into the corresponding phonetic transcription. Most methods use a lexicon, containing a large quantity of words or only syllables or tone groups. The creation of such a library is extremely complex and can be an individual implementation or a common lexicon used by several people. The quality can be continually improved by interactive user intervention. This means that users recognize defects in such a transcription formula. They improve their pronunciation manually and, gradually, their findings become an integral part of the lexicon.

The second step converts the phonetic transcription into an acoustic speech signal, where concatenation can be in the time or frequency range. While the first step normally has a software solution, the second step involves signal processors or dedicated processors.

In addition to the problems posed by coarticulation and prosody, speech recognition has to address pronunciation ambiguities to avoid generating misinterpretations like "the grass is full" or "the glass is fool" instead of the phrase "the glass is full." The only way to solve this problem is to provide additional information about the context.

3.7 Speech Input

Speech input deals with various applications, as shown in Figure 3-12.

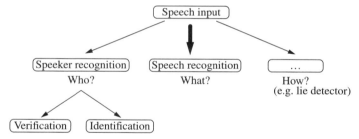

Figure 3-12 Speech input applications.

In the speech input context, we need to ask three questions to obtain correct answers: Who?, What?, and How?

- Who? Human speech has certain speaker-dependent characteristics, which means that speech input can serve to recognize a speaker [O'86]. The computer is used to recognize an acoustic fingerprint of the speaker. A typical application is personal identification, for example, in police departments. A speech signal is input into the system to identify the correct speaker. Another application is verification, for example, for access protection, where both the speech sample and the speaker are input. The system has to determine whether or not the speech sample belongs to the speaker.
- What? The central issue of speech input is to detect the speech contents themselves. A speech sequence is normally input to generate a piece of text. Typical

applications are speech-controlled typewriters, language translation systems, or accessibility options for users with special needs.

- How? Our third question relates to how a speech sample should be studied. One typical application is a lie detector.

3.7.1 Speech Recognition

Speech recognition is a very interesting field for multimedia systems. In combination with speech synthesis, it enables us to implement media transformations.

The primary quality characteristic of each speech recognition session is determined by a probability of ≤ 1 to recognize a word correctly. A word is always recognized only with a certain probability. Factors like environmental noise, room acoustics, and the physical and psychical state of the speaker play an important role. A poor recognition rate is $p=0.95$, which corresponds to five percent wrongly recognized words. With a sentence of only three words, the probability that the system will recognize all triples correctly drops to $p=0.05\times0.95\times0.95=0.86$.

This small example shows that a speech recognition system should have a very high single-word recognition rate. Figure 3-13 shows the conceptual components of such a system.

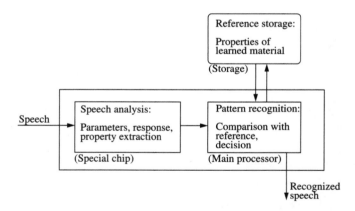

Figure 3-13 The speech recognition principle: the tasks are distributed to system components by the basic principle "extract characteristics to reduce data."

The speech recognition principle compares special characteristics of individual utterances with a sentence of previously extracted speech elements. This means that these characteristics are normally quantized for the speech sequence to be studied. The result is compared with existing references to allocate it to one of the existing speech elements. Identified utterances can be stored, transmitted, or processed as a parametrized sequence of speech elements.

Practical implementations normally use dedicated components or a signal processor to extract characteristic properties. The comparison and the decision are generally handled by the system's main processor, while the lexicon with reference characteristics normally resides in the computer's secondary storage unit.

Most practical methods differ in how they define characteristic properties. The principle shown in Figure 3-13 can be applied several times, each time referring to different characteristics. The application of the principle shown in Figure 3-13 can be divided into the steps shown in Figure 3-14.

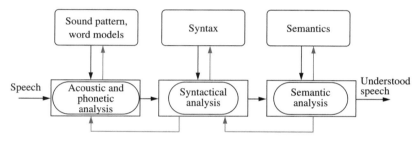

Figure 3-14 Speech recognition components.

The methods applied in the time and frequency ranges are:

1. Acoustic and phonetic analysis: Referring to the characteristic properties of the chosen method, the first step applies the principle shown in Figure 3-13 to sound patterns and/or word models.
2. Syntactic analysis: The second step uses the speech units determined in the first step to run a syntactic analysis on them. This process can detect errors in the first run. It serves as an additional decision tool because the first step does not normally provide a final decision.
3. Semantic analysis: The third step analyzes the semantics of the speech sequence recognized to this point. This step can detect errors from the previous decision process and remove them by using another interplay with other analytical methods. Note that even with current artificial intelligence and neural network technologies, the implementation of this step is extremely difficult.

These methods often work with characteristics in the time and/or frequency range. They are based on the same criteria and speech units (e.g., formants or phonemes) as in speech output.

A specific problem in speech input is the room acoustics, where environmental noise may prevail, so that frequency-dependent reflections overlay a sound wave along walls and objects with the primary sound wave. Also, word boundaries have to be defined, which is not easy, because most speakers or most human languages do not emphasize the end of one and the beginning of the next word. A kind of time standardization is required to be able to compare a speech unit with existing samples. The same word can be spoken fast or slow. However, we cannot simply clench or stretch the time axis, because elongation factors are not proportional to the total duration. There are long and short unvoiced sounds.

3.7.1.1 Speaker-Dependent and Speaker-Independent Speech Input Systems

We know that speaker recognition is the term applied to the recognition of a speaker based on his or her voice. Given equal reliability values, a speaker-dependent system can recognize many more words than a speaker-independent system, but at the cost of "training" the system in advance. To train a system to a speaker's speech, the speaker is normally requested to read specific speech sequences. Today, many speech recognition systems have a training phase of less than half an hour. Most speaker-dependent systems can recognize 25,000 words and more. The "hit" rate of a speaker-independent system is approximately 1,000 words. Note that these values are only rough indicators. A real-world system evaluation should include environmental factors, for example, whether the measurement was taken in an anechoic room.

3.8 Speech Transmission

Speech transmission is a field relating to highly efficient encoding of speech signals to enable low-rate data transmission, while minimizing noticeable quality losses. The following sections provide a short introduction to some important principles that interest us at the moment in connection with speech input and output. Encoding methods and audio compression are described in Chapter 7.

3.8.1 Pulse Code Modulation

Signal form encoding does not consider speech-dependent properties or parameters. The technologies applied are merely expected to offer efficient encoding of audio signals. A straightforward technique for digitizing an analog signal (waveform) is Pulse Code Modulation (PCM). This method is simple, but it still meets the high quality demands stereo-audio signals in the data rate used for CDs:

$$\text{rate} = 2 \times \frac{44{,}100}{s} \times \frac{16 \text{ bits}}{8 \text{bits/byte}} = 176{,}400 \text{ bytes/s}$$

As a side note, telephone quality requires only 64 Kbit/s compared to 176,400 byte/s for the case studied here. Differential Pulse Code Modulation (DPCM) achieves 56 Kbit/s in at least equal quality, while Adaptive Pulse Code Modulation (ADPCM) enables a further reduction to 32 Kbit/s.

3.8.2 Source Encoding

An alternative method is source encoding, where some transformations depend on the original signal type. For example, an audio signal has certain characteristics that can be exploited in compression. The suppression of silence in speech sequences is a typical example of a transformation that depends entirely on the signal's semantics.

Parametric systems use source encoding. They utilize speech-specific characteristics to reduce data, for example the channel vocoder shown in Figure 3-15.

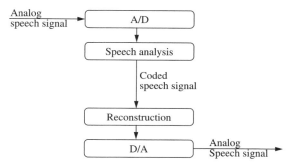

Figure 3-15 Components of a speech transmission system using source encoding.

A vocoder is an electronic mechanism that reduces speech signals to slowly varying signals that can be transmitted over communication systems of limited frequency bandwidth. A channel vocoder uses an enhanced subband encoding method. It analyzes speech by dividing the signal into a set of frequency bands, assuming that certain frequency maxima are relevant for speech and others are not. Figure 3-16 shows such a bandwidth division. In addition, the technique utilizes differences between voiced and unvoiced sounds. Unvoiced sounds are generated by means of a noise generator. A pulse sequence is selected to generate voiced sounds. These pulses have rates that correspond exactly to the basic frequency of the measured speech. However, the quality is not always satisfactory.

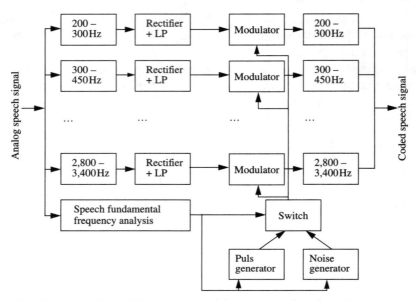

Figure 3-16 The speech analysis components of a channel vocoder.

3.8.3 Recognition-Synthesis Methods

Current research work attempts to further reduce the data volume by approximately 6 Kbit/s. The quality should always correspond to an uncompressed 64-Kbit/s signal. Experts also study ways to reduce the transmission rate of speech signals by use of pure recognition-synthesis methods (see Figure 3-17).

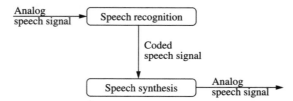

Figure 3-17 Components of a recognition-synthesis system for speech transmission.

This method conducts a speech analysis and a speech synthesis during reconstruction, offering a reduction to approximately 50 bit/s. Only the speech element characteristics are transmitted, for example formants containing data about the center frequencies and bandwidths for use by digital filters. However, the quality of the reproduced speech and its recognition rate need to be further improved to be acceptable.

3.8.4 Achievable Quality

One of the most important aspects of speech and audio transmission in multimedia systems is the minimal achievable data rate in a defined quality. An interesting discussion of this aspect is given in [Fla92] and shown in Figure 3-18. A data rate of less than 8 Kbit/s for telephone quality can be achieved.

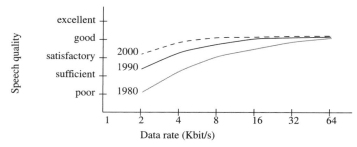

Figure 3-18 Quality of compressed speech in relation to the compressed signal's data rate.

Figure 3-18 relates the audio quality to the number of bits per sampling value. This ratio provides an excellent CD quality at a reduction of 16 bits per sampling value to 2 bits per sampling value, which means that only one eighth of the actual data rate is required to achieve this quality.

CHAPTER 4

Graphics and Images

Graphics and images are both non-textual information that can be displayed and printed. They may appear on screens as well as on printers but cannot be displayed with devices only capable of handling characters. This chapter discusses computerized graphics and images, their respective properties, and how they can be acquired, manipulated, and output on computers. This introductory discussion includes options to represent and process graphics and images on computers, some important formats, and automatic content analysis. (Chapter 9 describes how picture and image contents are processed.) Building on this introduction, we explain methods used to retransform two-dimensional images into three-dimensional space. Finally, the chapter presents particular techniques of how to output graphics and images on output devices (printers, display units). [EF94, FDFH92] provide a further discussion of digital image processing.

4.1 Introduction

Graphics are normally created in a graphics application and internally represented as an assemblage of objects such as lines, curves, or circles. Attributes such as style, width, and color define the appearance of graphics. We say that the representation is aware of the semantic contents. The objects graphics are composed of can be individually deleted, added, moved, or modified later. In contrast, images can be from the real world or virtual and are not editable in the sense given above. They ignore the semantic contents. They are described as spatial arrays of values. The smallest addressable image element is called a pixel. The array, and thus the set of pixels, is called a bitmap. Object-based editing is not possible, but image editing tools exist for enhancing and retouching bitmap images. The drawback of bitmaps is that they need much more storage capacity then graphics. Their advantage is that no processing is necessary before displaying them, unlike graphics where the abstract definition must be processed first to produce a bitmap. Of course, images captured from an analog signal, via scanners or video cameras, are represented as bitmaps, unless semantic recognition takes place such as in optical character recognition.

4.2 Capturing Graphics and Images

The process of capturing digital images depends initially upon the image's origin, that is, real-world pictures or digital images. A digital image consists of N lines with M pixels each.

4.2.1 Capturing Real-World Images

A picture is a two-dimensional image captured from a real-world scene that represents a momentary event from the three-dimensional spatial world. Figure 4-1 shows the camera obscura model of an image capturing system with focal length F, where the spatial Cartesian world coordinates $[W_1, W_2, W_3]$ specify the distance of a spatial point from the camera lens (the coordinate system's origin). These points are mapped onto the coordinates of the image level $w=[r,s]$ by applying the central projection equation:

$$r = F \times \frac{W_1}{W_3}; \quad s = F \times \frac{W_2}{W_3}$$

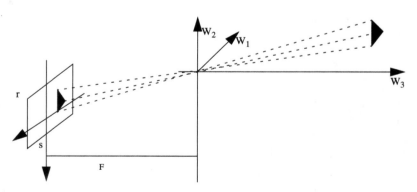

Figure 4-1 Projection of the real world onto the image plane.

An image capturing device, such as a CCD scanner or CCD camera for still images, or a frame grabber for moving images, converts the brightness signal into an electrical signal. In contrast to conventional TV standards that use a line structure system (see Chapter 5), the line direction of the output signal that the capturing device generates is normally continuous in the row direction, but discrete and analogous in the column direction. The first step in processing real-world pictures is to sample and digitizes these signals. The second step normally involves quantization to achieve an aggregation of color regions to reduce the number of colors, depending on the hardware used to output the images. Video technologies normally work with an 8-bit PCM quantization, which means they can represent $2^8=256$ different colors or gray levels per pixel. This results in $2^8 \times 2^8 \times 2^8$ or approximately 16 million different colors.

Next, the digitized picture is represented by a matrix composed of rows and columns to accommodate numerical values. Each matrix entry corresponds to a brightness value. If I specifies a two-dimensional matrix, then $I(r,c)$ is the brightness value at the position corresponding to row r and column c of the matrix.

The spatial two-dimensional matrix representing an image is made up of pixels—the smallest image resolution elements. Each pixel has a numerical value, that is, the number of bits available to code a pixel—also called amplitude depth or pixel depth. A numerical value may represent either a black (numerical value 0) or a white (numerical value 1) dot in bitonal (binary) images, or a level of gray in continuous-tone monochromatic images, or the color attributes of the picture element in color pictures. Numerical values for gray levels range from 0 for black to FF for white. Figure 4-2 shows an example with different tones.

2 gray levels   4 gray levels   256 gray levels

Figure 4-2 Images with different numbers of gray levels.

A rectangular matrix is normally used to represent images. The pixels of an image are equally distributed in the matrix, and the distance between the matrix dots is obviously a measure of the original picture's quality. It also determines the degree of detail and image's resolution, but the resolution of an image also depends on the representation system.

Digital images are normally very large. If we were to sample and quantize a standard TV picture (525 lines) by use of a VGA (Video Graphics Array; see Chapter 5) video controller in a way to be able to represent it again without noticeable deterioration, we would have to use a matrix of at least 640×480 pixels, where each pixel is represented by an 8-bit integer, allowing a total of 256 discrete gray levels. This image specification results in a matrix containing 307,200 eight-bit numbers, that is, a total of 2,457,600 bits. In many cases, the sampling would be more complex. So the question is how to store such high-volume pictures. The next section deals with image formats because they influence storage requirements of images. Later we describe image storage options.

4.2.2 Image Formats

The literature describes many different image formats and distinguishes normally between image capturing and image storage formats, that is, the format in which the image is created during the digitizing process and the format in which images are stored (and often transmitted).

4.2.2.1 Image Capturing Formats

The format of an image is defined by two parameters: the spatial resolution, indicated in pixels, and the color encoding, measured in bits per pixel. The values of both parameters depend on the hardware and software used to input and output images.

4.2.2.2 Image Storage Formats

To store an image, the image is represented in a two-dimensional matrix, in which each value corresponds to the data associated with one image pixel. In bitmaps, these values are binary numbers. In color images, the values can be one of the following:

- Three numbers that normally specify the intensity of the red, green, and blue components.
- Three numbers representing references to a table that contains the red, green, and blue intensities.
- A single number that works as a reference to a table containing color triples.
- An index pointing to another set of data structures, which represents colors.

Assuming that there is sufficient memory available, an image can be stored in uncompressed RGB triples. If storage space is scarce, images should be compressed in a suitable way (Chapter 7 describes compression methods). When storing an image, information about each pixel, i.e., the value of each color channel in each pixel, has to be stored. Additional information may be associated to the image as a whole, such as width and height, depth, or the name of the person who created the image. The necessity to store such image properties led to a number of flexible formats, such as RIFF (Resource Interchange File Format), or BRIM (derived from RIFF) [Mei83], which are often used in database systems. RIFF includes formats for bitmaps, vector drawings, animation, audio, and video. In BRIM, an image consists of width, height, authoring information, and a history field specifying the generation process or modifications.

The most popular image storing formats include PostScript, GIF (Graphics Interchange Format), XBM (X11 Bitmap), JPEG, TIFF (Tagged Image File Format), PBM

(Portable Bitmap), and BMP (Bitmap). The following sections provide a brief introduction to these formats; JPEG will be described in Chapter 7.

4.2.2.3 PostScript

PostScript is a fully fledged programming language optimized for printing graphics and text (whether on paper, film, or CRT). It was introduced by Adobe in 1985. The main purpose of PostScript was to provide a convenient language in which to describe images in a device-independent manner. This device independence means that the image is described without reference to any specific device features (e.g., printer resolution) so that the same description could be used on any PostScript printer without modification. In practice, some PostScript files do make assumptions about the target device (such as its resolution or the number of paper trays it has), but this is bad practice and limits portability. During its lifetime, PostScript has been developed in levels, the most recent being Level 3:

- Level 1 PostScript: The first generation was designed mainly as a page description language, introducing the concept of scalable fonts. A font was available in either 10 or 12 points, but not in an arbitrary intermediate size. This format was the first to allow high-quality font scaling. This so-called Adobe Type 1 font format is described in [Inc90].

- Level 2 PostScript: In contrast to the first generation, Level 2 PostScript made a huge step forward as it allowed filling of patterns and regions, though normally unnoticed by the nonexpert user. The improvements of this generation include better control of free storage areas in the interpreter, larger number of graphics primitives, more efficient text processing, and a complete color concept for both device-dependent and device-independent color management.

- Level 3 PostScript: Level 3 takes the PostScript standard beyond a page description language into a fully optimized printing system that addresses the broad range of new requirements in today's increasingly complex and distributed printing environments. It expands the previous generation's advanced features for modern digital document processing, as document creators draw on a variety of sources and increasingly rely on color to convey their messages.

At some point, you may want to include some nice PostScript images into a document. There are a number of problems associated with this, but the main one is that your page layout program needs to know how big the image is and how to move it to the correct place on the page. Encapsulated PostScript (EPS) is that part of Adobe's Document Structuring Convention (DSC) that provides this information. An EPS file is a PostScript file that follows the DSC and that follows a couple of other rules. In contrast to Postscript, the EPS format has some drawbacks:

- EPS files contain only one image.
- EPS files always start with comment lines, e.g., specifying the author and resources (e.g., fonts).

Detailed information on the EPS format is contained in [Sch97a]. The PostScript language and DSC specifications are available at www.adobe.com.

4.2.2.4 Graphics Interchange Format (GIF)

The Graphics Interchange Format (GIF) was developed by CompuServe Information Service in 1987. Three variations of the GIF format are in use. The original specification, GIF87a, became a de facto standard because of its many advantages over other formats. Creators of drawing programs quickly discovered how easy it was to write a program that decodes and displays GIF images. GIF images are compressed to 20 to 25 percent of their original size with no loss in image quality using a compression algorithm called LZW (see Chapter 7). The next update to the format was the GIF89a specification. GIF89a added some useful features, including transparent GIFs.

Unlike the original GIF specifications, which support only 256 colors, the GIF24 update supports true 24-bit colors, which enables you to use more than 16 million colors. One drawback to using 24-bit color is that, before a 24-bit image can be displayed on an 8-bit screen, it must be dithered, which requires processing time and may also distort the image. GIF24 uses a compression technique called PNG.

4.2.2.5 Tagged Image File Format (TIFF)

The Tagged Image File Format (TIFF) was designed by Aldus Corporation and Microsoft [Cor92] in 1987 to allow portability and hardware independence for image encoding. It has become a de facto standard format. It can save images in an almost infinite number of variations. As a result, no available image application can claim to support all TIF/TIFF file variations, but most support a large number of variations.

TIFF documents consist of two components. The baseline part describes the properties that should support display programs. The second part are extensions used to define properties, that is, the use of the CMYK color model to represent print colors.

An important basis to be able to exchange images is whether or not a format supports various color models. TIFF offers binary levels, gray levels, palettes, RGB, and CMYK colors. Whether or not an application supports the color system specified in TIFF extensions depends on the respective implementation.

TIFF supports a broad range of compression methods, including run-length encoding (which is called PackBits compression in TIFF jargon), LZW compression, FAX Groups 3 and 4, and JPEG (see Chapter 7). In addition, various encoding methods, including Huffman encoding, can be used to reduce the image size.

TIFF differs from other image formats in its generics. In general, the TIFF format can be used to encode graphical contents in different ways, for example, to provide previews (thumbnails) for quick review of images in image archives without the need to open the image file.

4.2.2.6 X11 Bitmap (XBM) and X11 Pixmap (XPM)

X11 Bitmap (XBM) and X11 Pixmap (XPM) are graphic formats frequently used in the UNIX world to store program icons or background images. These formats allow the definition of monochrome (XBM) or color (XPM) images inside a program code. The two formats use no compression for image storage.

In the monochrome XBM format, the pixels of an image are encoded and written to a list of byte values (byte array) in the C programming language, grouping 8 pixels into a byte value. There are two additional definitions for the dimensions of an image (see Figure 4-3).

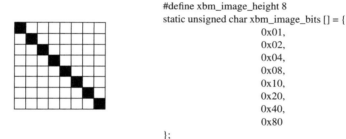

Figure 4-3 Example of an XBM image.

In XPM format, image data are encoded and written to a list of strings (string array), together with a header. The first line defines the image dimension and a hot spot, which is used as a cursor icon to identify the exact pixel that triggers a mouse selection. The next lines describe the colors used, replacing a text or an RGB color value by a character from the ASCII character set. The lines that follow these lines list the image lines, including their replaced color values (see Figure 4-4). Note that unused pixels are represented by blanks.

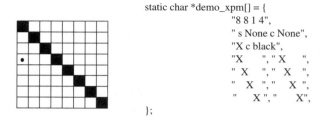

Figure 4-4 Example of an XPM image.

Neither XBM nor XPM images are compressed for storage. This means that their representation by 8-bit ASCII values generates always the same data volume. Both formats allow encoding of only 256 colors or gray levels.

4.2.2.7 Portable Bitmap plus (PBMplus)

PBMplus is a software package that allows conversion of images between various image formats and their script-based modification. PBMplus includes four different image formats, Portable Bitmap (PBM) for binary images, Portable Graymap (PGM) for gray-value images, Portable Pixmap (PPM) for true-color images, and Portable Anymap (PNM) for format-independent manipulation of images. These formats support both text and binary encoding. The software package contains conversion tools for internal graphic formats and other formats, so that it offers free and flexible conversion options.

The contents of PBMplus files are the following:

- A magic number identifying the file type (PBM, PGM, PPM, or PNM), that is, "P1" for PBM.
- Blanks, tabs, carriage returns, and line feeds.
- Decimal ASCII characters that define the image width.
- Decimal ASCII characters that define the image height.
- ASCII numbers plus blanks that specify the maximum value of color components and additional color information (for PPM, PNM, and PBM).

Filter tools are used to manipulate internal image formats. The functions offered by these tools include color reduction; quantization and analysis of color values; modification of contrast, brightness and chrominance; cutting, pasting and merging of several images; changing the size of images; or generating textures and fractal backgrounds.

4.2.2.8 Bitmap (BMP)

BMP files are device-independent bitmap files most frequently used in Windows systems. The BMP format is based on the RGB color model. BMP does not compress the original image. The BMP format defines a header and a data region. The header region (BITMAPINFO) contains information about size, color depth, color table, and compression method. The data region contains the value of each pixel in a line. Lines are flush-extended to a value divisible by 32 and padded with zero values.

Valid color depth values are 1, 4, 8, and 24. The BMP format uses the run-length encoding algorithm to compress images with a color depth of 4 or 8 bits/pixel, where two bytes each are handled as an information unit. If the first byte value contains zero and the second value is greater than three, then the second value contains the number of bytes that follow and contains the color of the next pixel as a reference to the color table (no compression). Otherwise, the first byte value specifies the number of pixels that follow, which are to be replaced by the color of the second byte value to point to the color table. An image encoded with 4 bits/pixel uses only four bits for this information. In the header region, BMP defines an additional option to specify a color table to be used to select colors when the image is displayed.

4.2.3 Creating Graphics

4.2.3.1 Input Devices

Modern graphical input devices include mice (with or without cables), tablets, and transparent, highly sensitive screens, or input devices that allow three-dimensional or higher-dimensional input values (degrees of freedom), in addition to the x and y positions on the screen, such as trackballs, spaceballs, or data gloves.

Some trackball models available today rotate around the vertical axis in addition to the two horizontal axes, but there is no direct relationship between hand movements on the device and the corresponding movement in three-dimensional space.

A spaceball is a solid ball positioned on an elastic base. Pressing or pulling the ball in any direction produces a 3D translation and a 3D orientation. The directions of movement correspond to the user's attempts to move the solid ball although the hand does not actually move.

A data glove is a device that senses the hand position and its orientation [ZLB[+]87]. It allows pointing to objects, exploring and navigating within scenes, or acting at a distance on the real world. Virtual objects may be manipulated, for example rotated for further examination. With gloves, real objects may be moved at a distance, while the user only monitors their virtual representation. High-precision gloves are sophisticated and expensive instruments. Special gloves may feed back tactile sensations by means of "tactile corpuscles," which exert pressure on the finger tips. Shapes of objects may thus be simulated. Research is also going on in the simulation of object textures.

4.2.3.2 Graphics Software

Graphics are generated by use of interactive graphic systems. The conceptual environment of almost all interactive graphic systems is an aggregated view consisting of three software components—application model, application program, and graphics system—and one hardware component.

The application model represents data or objects to be displayed on the screen. It is normally stored in an application database. The model acquires descriptions of primitives that describe the form of an object's components, attributes, and relations that explain how the components relate to each other. The model is specific to an application and independent of a system used for display. This means that the application program has to convert a description of parts of the model into procedure calls or commands the graphic system can understand to create images. This conversion process is composed of two phases. First, the application program searches the application database for parts to be considered, applying certain selection or search criteria. Second, the extracted geometry is brought into a format that can be passed on to the graphics system.

The application program processes user input and produces views by sending a series of graphical output commands to the third component, the graphics system. These output commands include a detailed geometric description as to what is to be viewed and how the objects should appear.

The graphics system is responsible for image production involving detailed descriptions and for passing user input on to the application program (for processing purposes). Similar to an operating system, the graphics system represents an intermediate component between the application program and the display hardware. It influences the output transformation of objects of the application model into the model's view. In a symmetric way, it also influences the input transformation of user actions for application program inputs leading to changes in the model and/or image. A graphics system normally consists of a set of output routines corresponding to various primitives, attributes, and other elements. The application program passes geometric primitives and attributes on to these routines. Subroutines control specific output devices and cause them to represent an image.

Interactive graphics systems are an integral part of distributed multimedia systems. The application model and the application program can represent applications as well as user interfaces. The graphics system uses (and defines) programming abstractions supported by the operating system to establish a connection to the graphics hardware.

4.2.4 Storing Graphics

Graphics primitives and their attributes are on a higher image representation level because they are generally not specified by a pixel matrix. This higher level has to be mapped to a lower level at one point during image processing, for example, when representing an image. Having the primitives on a higher level is an advantage, because it reduces the data volume that has to be stored for each image and allows simpler image manipulation. A drawback is that an additional step is required to convert graphics primitives and their attributes into a pixel representation. Some graphics packages, for example, the SRGP (Simple Raster Graphics Package), include this type of conversion. This means that such packages generate either a bitmap or a pixmap from graphics primitives and their attributes.

We have seen that a bitmap is a pixel list that can be mapped one-to-one to pixel screens. Pixel information is stored in 1 bit, resulting in a binary image that consists exclusively of black and white. The term pixmap is a more general description of an image that uses several bits for each pixel. Many color systems use 8 bits per pixel (e.g., GIF), so that 256 colors can be represented simultaneously. Other formats (including JPEG) allow 24 bits per pixel, representing approximately 16 million colors.

Other packages—for example, PHIGS (Programmer's Hierarchical Interactive Graphics System) and GKS (Graphical Kernel System)—use graphics specified by primitives and attributes in pixmap form [FDFH92].

4.2.4.1 Graphics Storage Formats

File formats for vector graphics allow loading and storing of graphics in a vectored representation, such as files created in a vector graphics application. The most important file formats include:

- IGES: The Initial Graphics Exchange Standard was developed by an industry committee to formulate a standard for the transfer of 2D and 3D CAD data.
- DXF: AutoDesk's 2D and 3D format was initially developed for AutoCAD, a computer-aided design application. It has become a de facto standard.
- HPGL: The Hewlett Packard Graphics Language has been designed to address plotters, which is the reason why it only supports 2D representation.

The combination of vector and raster graphics is generally possible in modern vector graphics systems. With regard to representing data in files, the two graphics types are often totally separated from one another. Only a few so-called meta file formats—for example, CGM (Computer Graphics Metafile), PICT (Apple Macintosh Picture Format), and WMF (Windows Metafile)—allow an arbitrary mixture of vector and raster graphics.

4.3 Computer-Assisted Graphics and Image Processing

Computer graphics deal with the graphical synthesis of real or imaginary images from computer-based models. In contrast to this technique, image processing involves the opposite process, that is, the analysis of scenes, or the reconstruction of models from images representing 2D or 3D objects. The following sections describe some image analysis (image recognition) and image synthesis (image generation) basics. For detailed information, see [FDFH92, KR82, Nev82, HS92, GW93].

4.3.1 Image Analysis

Image analysis involves techniques to extract descriptions from images, which are required by methods used to analyze scenes on a higher level. Knowing the position and the value of a particular pixel does not contribute enough information to recognize an object, to describe the object form and its position or orientation, to measure the distance to an object, or whether or not an object is defective. Techniques applied to analyzing images include the calculation of perceived colors and brightness, a partial or full reconstruction of three-dimensional data in a scene, and the characterization of the properties of uniform image regions.

Image analysis is important in many different fields, for example, to evaluate photos taken by an air contamination monitor, sampled TV pictures of the moon or other planets received from space probes, TV pictures generated by the visual sensor of an industrial robot, X-ray pictures, or CAT (Computerized Axial Tomography) pictures. Some image processing fields include image improvement, pattern discovery and recognition, scene analysis, and computer vision.

Image improvement is a technique to improve the image quality by eliminating noise (due to external effects or missing pixels), or by increasing the contrast.

Pattern discovery and pattern recognition involve the discovery and classification of standard patterns and the identification of deviations from these patterns. An important example is OCR (Optical Character Recognition) technology, which allows efficient reading of print media, typed pages, or handwritten pages into a computer. The degree of accuracy depends on the source material and the input device, particularly when reading handwritten pages. Some implementations allow users to enter characters by use of a steadily positioned device, normally a tablet pen, which are detected by the computer (online character recognition). This process is much easier compared to the recognition of scanned characters because the tablet acquires sequence, direction, and—in some cases—speed and pressure. Also, a pattern recognition algorithm can match these factors for each character against stored templates. The recognition process can evaluate patterns without knowing how the patterns have been created (static pattern recognition). Alternatively, it can respond to pressure, character edges, or the drawing speed (dynamic pattern recognition). For example, a recognition process can be trained to recognize various block capital styles. In such a process, the parameters of each character are calculated from patterns provided by the user. [KW93] describes an architecture of an object-oriented character recognition engine (AQUIRE), which supports online recognition with combined static and dynamic capabilities.

Scene analysis and computer vision concern the recognition and reconstruction of 3D models of a scene consisting of various 2D images. A practical example is an industrial robot that measures the relative sizes, shapes, positions, and colors of objects.

The following sections identify some properties that play an important role in recognizing images. Information yielded by these methods is subsequently aggregated to allow recognition of image contents.

4.3.1.1 Image Properties

Most image recognition methods use color, texture, and edges to classify images. Once we have described an image based on these three properties, we can, for example, query an image database by telling the system, "Find an image with a texture similar to that of the sample image."

Color One of the most intuitive and important characteristics to describe images is color. Assume that the image we want to analyze is available in the usual RGB format with three 8-bit color channels. The basic approach is to use a color histogram to acquire the image, that is, how many pixels of the image take a specific color. To avoid working with a indeterminable number of colors, we previously discretize the colors occurring in the image. We achieve this by using only the n leftmost bits of each channel. With $n=2$, our histogram will have entries for 64 colors. Figure 4-5 shows an example of a gray-value histogram (because this book is printed in black on white paper) for an image with a palette of 256 possible gray values.

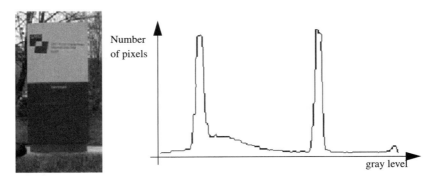

Figure 4-5 The gray-value histogram of an image.

For humans to recognize the colorfulness of an image, it is relevant whether or not a color tends to occur on large surfaces (in a coherent environment) or in many small spots, in addition to the frequency in which a color occurs. This information will be lost if we were to count only the number of pixels in a specific color. For this reason, [PZM96] suggests a so-called color coherence vector (CCV). To calculate CCV, each pixel is checked as to whether it is within a sufficiently large one-color environment (i.e., in a region related by a path of pixels in the same color). If so, we call it coherent, otherwise it is incoherent. We prepare two separate histograms to count coherent and incoherent pixels for each color.

Assume we determined J colors after the discretization process, then α_j or β_j ($j=1,...,J$) describe the number of coherent or incoherent pixels of the color j. The color coherence vector is then given by $((\alpha_1,\beta_1),...,(\alpha_J,\beta_J))$, which we store to describe the colorfulness of the image. When comparing two images, B and B', against CCVs $((\alpha_1,\beta_1),...,(\alpha_J,\beta_J))$ or $((\alpha_1',\beta_1'),...,(\alpha_J',\beta_J'))$, we use the expression

$$dist(B, B') = \sum_{j=1}^{J}\left(\left|\frac{\alpha_j - \alpha_j'}{\alpha_j + \alpha_j' + 1}\right| + \left|\frac{\beta_j - \beta_j'}{\beta_j + \beta_j' + 1}\right|\right)$$

as a measure of distance.

An advantage of using color as our property to compare two images is that it is robust and can represent slight changes in scaling or perspective and it allows fast calculation. However, we cannot normally use the Euclidean distance of two color vectors to draw a direct conclusion on the difference in human color recognition. We can solve this problem by transforming the RGB image before we transform the histogram into a color space, which corresponds better to human recognition. One such color space is the so-called L*a*b* space (see [Sch97a]).

Texture A texture is a small surface structure, either natural or artificial, regular or irregular. Examples of textures are wood barks or veining, knitting patterns, or the surface of a sponge, as shown in Figure 4-6. When studying textures, we distinguish between two basic approaches. First, the structural analysis searches for small basic components and an arrangement rule, by which to group these components to form a texture. Second, the statistical texture analysis describes the texture as a whole based on specific attributes, for example, local gray-level variance, regularity, coarseness, orientation, and contrast. These attributes are measured in the spatial domain or in the spatial frequency domain, without decoding the texture's individual components. In practical applications, structural methods do not play an important role.

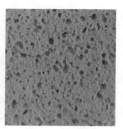

Figure 4-6 Some texture examples.

To analyze textures, color images are first converted into a gray-level representation. When studying natural images (e.g., landscape photographs), we have to deal with the issue of what structures we want to call a texture (which depends on the scaling and other factors) and where in the image there may be textured regions. To solve this issue, we can use a significant and regular variation of the gray values in a small environment [KJB96] as a criterion for the occurrence of a texture in an image region. Once we have opted for a texture measuring unit, we determine such homogeneous regions in the image segmentation process. Finally, we calculate the texture measuring unit for each texture region.

To illustrate this process, we use a simplified statistical method to analyze a texture in the local space. We calculate and interpret gray-level co-occurrence matrices [Zam89]. These matrices state how often two gray values, a and b, occur in an image in a specific arrangement. For an arrangement, $[a][b]$ (i.e., gray-level b is immediately right of pixels with gray-level a), Figure 4-7 shows the gray-level co-occurrence matrix on the right for the sample represented on the left.

0	0	0	2	2
0	0	0	2	2
0	0	0	3	3
1	1	1	3	3
1	1	1	3	3

b \ a	0	1	2	3
0	6	0	0	0
1	0	4	0	0
2	2	0	2	0
3	1	2	0	3

Figure 4-7 Example of a gray-level co-occurrence matrix.

We could form gray-level co-occurrence matrices for any other neighboring arrangement of two gray values. If we distinguish N gray values and call the entries in an $N \times N$ gray-level co-occurrence matrix $g(a,b)$, then

$$K = \sum_{a=0}^{N-1} \sum_{b=0}^{N-1} (a-b)^2 g(a,b)$$

can be used as a measurement unit for a texture's contrast. Intuitively, we speak of high contrast when there are very different gray values in a dense neighborhood. If the gray levels are very different, $(a-b)^2$, and if they border each other frequently, then $g(a,b)$ takes a high value [Zam89]. It is meaningful not to limit to one single neighboring arrangement. The expression

$$H = \sum_{a=0}^{N-1} \sum_{b=0}^{N-1} g(a,b)^2$$

measures a texture's homogeneity, because in a homogeneous and perfectly regular texture, there are only very few different gray-level co-occurrence arrangements (i.e., essentially only those that occur in the small basic component), but they occur frequently.

Powerful texture analysis methods are based on multiscale simultaneous autoregression, Markov random fields, and tree-structured wavelet transforms. These methods go beyond the scope of this book; they are described in detail in [PM95, Pic96].

Edges The use of edges to classify images provides a basic method for image analysis—the convolution of an image from a mask [Hab95]. This method uses a given input image, E, to gradually calculate a (zero-initialized) output image, A. A convolution mask (also called convolution kernel), M, runs across E pixel by pixel and links the entries in the mask at each position that M occupies in E with the gray value of the underlying image dots. The result of this linkage (and the subsequent sum across all products from the mask entry and the gray value of the underlying image pixel) is written to output image A. The terms

- $e(x, y)$: gray value of the pixel at position (x, y) in input image E
- $a(x, y)$: entry at position (x, y) in output image A
- m: size of mask M, that is, $m \times m$, m uneven
- $m(u, v)$: entry at position (u, v); $u, v = 0, ..., m-1$ in mask M

are used to calculate $k=(m-1)/2$ from:

$$a(x, y) = \sum_{u, v = 0}^{m-1} e(x + k - u, y + k - v) m(u, v)$$

where marginal areas of width k remain initial in A. Figure 4-8 shows this method.

6	1	3	8	7	8	...	...
6	2	5	7	7	8	...	...
6	1	4	8	9	9	...	...
...	...	...	...	...	...	...	...
...	...	...	...	...	...	...	...

E

1	0	-1
2	0	-2
1	0	-1

M

7	-24	-13	-3	...	...	...
...	...	...	...	...	...	...
...	...	...	...	...	...	...
...	...	...	...	...	...	...

A

Figure 4-8 Using a filter.

This method is applied in the edge extraction process. A preliminary step smoothes the image available in gray-level representation, where the gray value of each pixel is replaced by a suitably weighed mean value of the gray values of its neighboring pixels, at the cost of losing some sharpness. However, this loss is tolerable because the method yields a less noisy image.

To locate pixels along the vertical edges (that is, pixels experiencing a strong horizontal gray-level transition), a mask, M_{horiz} as shown in Figure 4-9, runs across the smoothed image. The Sobel operators used in this example are particularly suitable to locate gray-value transitions. In output image A_{horiz}, a high-quantity entry means that there is a significant change in gray-levels in the "left-right direction" at the relevant position in E. The sign specifies the transition's direction. The approach for the horizontal edges is similar; the convolution of E with M_{vert} results in A_{vert}.

$$M_{horiz} = \begin{bmatrix} 1 & 0 & -1 \\ 2 & 0 & -2 \\ 1 & 0 & -1 \end{bmatrix} \qquad M_{vert} = \begin{bmatrix} 1 & 2 & 1 \\ 0 & 0 & 0 \\ -1 & -2 & -1 \end{bmatrix}$$

Figure 4-9 Using Sobel operators.

This concept provides partial derivations from E in the column and line directions. To determine the gradient amount that specifies the total strength of an "oblique" edge at position (x,y), we determine

$$a_{grad}(x, y) = \sqrt{a_{horiz}(x, y)^2 + a_{vert}(x, y)^2}$$

and finally binarize the result by use of a threshold value, E. This means that only the pixels in which a sufficient gray-value gradient was determined will flow into the final output image. During binarization, 1s are entered for all pixels that exceed the threshold value and 0s are entered for all remaining pixels in the output image.

Subsequent steps involve calculation of the gradient orientation and determination of the quantities of pixels that are on the same edge.

4.3.1.2 Image Segmentation

Segmentation is an operation that assigns unique numbers (identifiers) to object pixels based on different intensities or colors in the foreground and background regions of an image. A zero is assigned to the background pixels. Segmentation is primarily used to identify related pixel areas, so-called objects, while the recognition of these objects is not part of the segmentation process.

The following example uses gray-level images, but the methods can be easily applied to color images by studying the R, G, and B components separately.

Segmentation methods are classified as follows [Bow92, Jäh97, Jai89, RB93, KR82]:

- Pixel-oriented methods
- Edge-oriented methods
- Region-oriented methods

In addition to these methods, many other methods are known [GW93, Fis97b], for example, methods working with neuronal networks.

Pixel-Oriented Segmentation Methods The segmentation criterion in pixel-oriented segmentation is the gray value of a pixel studied in isolation. This method acquires the gray-level distribution of an image in a histogram and attempts to find one or several thresholds in this histogram [Jäh97, LCP90]. If the object we search has one color but a different background, then the result is a bimodal histogram with spatially separated maxima. Ideally, the histogram has a region without pixels. By setting a threshold at this position, we can divide the histogram into several regions.

Alternatively, we could segment on the basis of color values. In this case, we would have to modify the method to the three-dimensional space we study so that pixel clouds can be separated. Of course, locating a threshold value in a one-dimensional space of gray values is much easier. In practical applications, this method has several drawbacks:

- Bimodal distribution hardly occurs in nature, and digitization errors prevents bimodal distribution in almost all cases [Jäh97].
- If the object and background histograms overlap, then the pixels of overlapping regions cannot be properly allocated.
- To function properly, these methods normally require that the images be edited manually. This step involves marking the regions that can be segmented by use of local histograms.

Edge-Oriented Segmentation Methods Edge-oriented segmentation methods work in two steps. First, the edges of an image are extracted, for example, by use of a Canny operator [Can86]. Second, the edges are connected so that they form closed contours around the objects to be extracted.

The literature describes several methods to connect edge segments into closed contours [GW93, KR82]. Many methods use the Hough transform to connect edges. This algorithm replaces a regression line by m predefined pixels, which means that the share of straight lines contained in an image can be determined in the image region.

The Hough transform is a technique that can be used to isolate features of a particular shape within an image. Because it requires that the desired features be specified in some parametric form, the classical Hough transform is most commonly used for the detection of regular curves such as lines, circles, ellipses, and so on. A generalized Hough transform can be employed in applications where a simple analytic description of features is not possible. Due to the computational complexity of the generalized Hough algorithm, we restrict the main focus of this discussion to the classical Hough transform. Despite its domain restrictions, the classical Hough transform retains many applications, as most manufactured parts (and many anatomical parts investigated in medical imagery) contain feature boundaries that can be described by regular curves. The main advantage of the Hough transform technique is that it is tolerant of gaps in feature boundary descriptions and is relatively unaffected by image noise.

The Hough technique is particularly useful for computing a global description of a feature (where the number of solution classes need not be known a priori), given (possibly noisy) local measurements.

As a simple example, Figure 4-10 shows a straight line and the pertaining Hough transform. We use the Hough transform to locate straight lines contained in the image. These lines provide hints as to how the edges we want to extract can be connected. To achieve this goal, we take the straight lines starting at one edge end in the Hough transform to the next edge segment into the resulting image to obtain the contours we are looking for.

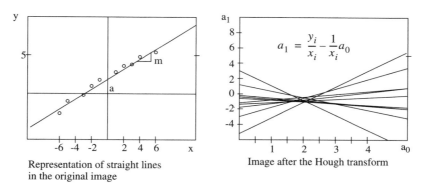

Figure 4-10 Using the Hough transform to segment an image.

Once we have found the contours, we can use a simple region-growing method to number the objects. The region-growing method is described in the following section.

Region-Oriented Segmentation Methods Geometric proximity plays an important role in object segmentation [CLP94]. Neighboring pixels normally have similar properties. This aspect is not taken into account in the pixel- and edge-oriented methods. In contrast, region-growing methods consider this aspect by starting with a "seed" pixel. Successively, the pixel's neighbors are included if they have some similarity to the seed pixel. The algorithm checks on each k level and for each region $R_i^k (1 \leq i \leq N)$ of the image's N regions whether or not there are unqualified pixels in the neighborhood of the marginal pixels. If an unqualified marginal pixel, x, is found, then the algorithm checks whether or not it is homogeneous to the region to be formed.

For this purpose, we use homogeneity condition $P(R_i^k \cup \{x\}) = \text{TRUE}$. For example, to check homogeneity P, we can use the standard deviation of the gray levels of one region.

Ideally, one seed pixel should be defined for each region. To automate the seed pixel setting process, we could use the pixels that represent the maxima from the histograms of the gray-level distribution of the images [KR82].

An efficient implementation of the required algorithm works recursively. Assume that a function, ELEMENT, is defined. The function specifies whether or not the homogeneity condition is met and one pixel has been marked. The steps involved in this algorithm are the following:

```
FUNCTION regionGrowing(x,y)
    if (ELEMENT(x,y))
       mark pixel as pertaining to object
    else
       if (ELEMENT(x-1,y-1)) regionGrowing(x-1,y-1)
       if (ELEMENT(x-1,y))   regionGrowing(x-1,y)
       if (ELEMENT(x-1,y+1)) regionGrowing(x-1,y+1)
       if (ELEMENT(x,y-1))   regionGrowing(x,y-1)
       if (ELEMENT(x,y+1))   regionGrowing(x,y+1)
       if (ELEMENT(x+1,y-1)) regionGrowing(x+1,y-1)
       if (ELEMENT(x+1,y))   regionGrowing(x+1,y)
       if (ELEMENT(x+1,y+1)) regionGrowing(x+1,y+1)
```

A drawback of this algorithm is that it requires a large stack region. If the image consists of one single object, we need a stack size corresponding to the image size. With a 100×100-pixel image, we would need a stack with a minimum depth of 10,000 pixels. Note that the stack depth of modern workstations is much smaller. For this reason, a less elegant iterative variant [FDFH92] is often used.

Region growing is based on a bottom-up approach, because smallest image regions grow gradually into larger objects. A drawback of this method is its high computing requirement. This problem can be solved by the split-and-merge algorithm—an efficient top-down algorithm [BD97, CMVM86, CP79], however at the cost of losing contour sharpness. The split step studies a square region as to homogeneity.

The gray-level mean value of a region is often used as a homogeneity criterion:

$$m_k = \frac{1}{n^2} \sum_{i=1}^{N} \sum_{j=1}^{N} P(i,j)$$

with standard deviation

$$\sigma_k = \sqrt{\frac{1}{n^2} \sum_{i=1}^{N} \sum_{j=1}^{N} (P(i,j) - m_k)^2}$$

where a region, k, is homogeneous with regard to a threshold, T, if $\sigma_k < T$.

Selecting threshold value T enables us to determine the granularity of the segmentation result. An inhomogeneous region is divided into four equally sized subregions and the process is repeated. However, this process may produce neighboring and homogeneous regions, which have to be merged in the subsequent merge step.

If ($|m_1 - m_2| < k \times \sigma_i, i = 1, 2$), we can merge two regions, where k serves as a factor specifying the granularity of the segmented image. Figure 4-11 shows this algorithm.

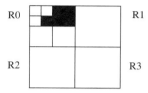

Figure 4-11 Using the split-and-merge algorithm.

Edge-oriented segmentation does not take interrelationships between regions into account during segmentation, that is, it utilizes only local information. The region-growing algorithm makes efficient use of pixel relationships, which means that it utilizes global information. On the other hand, it does not make efficient use of local edge information.

Ongoing work attempts to solve the discrepancy problem between the two approaches [Fis97a, Hen95]. The following section describes an algorithm that combines both approaches.

Water-Inflow Segmentation The previous sections described local and global segmentation strategies. A more efficient segmentation can be achieved by combining the local and global approaches [HB90, Hen95, PL90, Fis97a]. The example given in this section uses the so-called water-inflow algorithm [Fis97a].

The basic idea behind the water-inflow method is to fill a gray-level image gradually with water. During this process, the gray levels of the pixels are taken as height. The higher the water rises as the algorithm iterates, the more pixels are flooded. This means that land and water regions exist in each of the steps involved. The land regions correspond to objects. The second steps unifies the resulting segmented image parts into a single image. This method is an expansion of the watershed method introduced by Vincent and Soille [VS91]. Figure 4-12 shows the segmentation result at water level 30.

Original image

Segmented image

Figure 4-12 Segmentation result at a specific water level.

One advantage of this approach is that depth information lost when matching three-dimensional real-world images with two-dimensional images can be recovered, at least in part. Assume we look at a bent arm. Most segmentation strategies would find three objects: the upper part of the body, the arm, and the lower part of the body (which implies that the arm has a different gray level than the rest of the body). The water-inflow algorithm can recognize that the arm is in front of the body, from a spatial view. For this view, we need to create a hierarchy graph.

Now let's see how this algorithm works. First, it finds large objects, which are gradually split into smaller objects as the algorithm progresses. (The first object studied is the complete image.) This progress can be represented in a hierarchical graph. The root of the graph is the entire image. During each step, while objects are cloned, more new nodes are added to the graph. As new nodes are added, they become the sons of the respective object nodes. These nodes store the water level found for an object and a seed pixel for use in the region-growing process. This arrangement allows relocation of an arbitrary subobject when needed. The different gray levels of the subobjects ensure that the subobjects can be recognized as the water level rises and the algorithm progresses.

This means that, by splitting objects into smaller objects, we obtain an important tool to develop an efficient segmentation method. This aspect is also the main characteristic of this approach versus the methods described earlier.

4.3.1.3 Image Recognition

Previous sections described properties that can be used to classify the content of an image. So far, we have left out how image contents can be recognized. This section describes the entire process required for object recognition.

The complete process of recognizing objects in an image implies that we recognize a match between the sensorial projection (e.g., by a camera) and the observed image. How an object appears in an image depends on the spatial configuration of the pixel values. The following conditions have to be met for the observed spatial configuration and the expected projection to match:

- The position and the orientation of an object can be explicitly or implicitly derived from the spatial configuration.
- There is a way to verify that the derivation is correct.

To derive the position, orientation, and category or class of an object (e.g., a cup) from the spatial configuration of gray levels, we need a way to determine the pixels that are part of an object (as described in the segmentation section). Next, we need a way to distinguish various observed object characteristics from the pixels that are part of an object, for example, special markings, lines, curves, surfaces, or object boundaries (e.g., the rim of a cup). These characteristics are, in turn, organized in a spatial image-object relationship.

The analytical derivation of form, position, and orientation of an object depends on whether various object characteristics (a dot, line segment, or region in the two-dimensional space) match with corresponding object properties (a dot, line segment, circle segment, or a curved or plane surface).

The types of object, background, or sensor used to acquire the image, and the sensor's location determine whether a recognition problem is difficult or easy. Take, for example, a digital image with an object in the form of a white plane square on an evenly black background (see Table 4-1).

0	0	0	0	0	0	0	0	0	0	0	0	0
0	0	0	0	0	0	0	0	0	0	0	0	0
0	0	0	0	FF	FF	FF	FF	FF	0	0	0	0
0	0	0	0	FF	FF	FF	FF	FF	0	0	0	0
0	0	0	0	FF	FF	FF	FF	FF	0	0	0	0
0	0	0	0	FF	FF	FF	FF	FF	0	0	0	0
0	0	0	0	FF	FF	FF	FF	FF	0	0	0	0
0	0	0	0	0	0	0	0	0	0	0	0	0
0	0	0	0	0	0	0	0	0	0	0	0	0

Table 4-1 Numeric digital intensity image with a white square (gray level FF) on black background (gray level 0) of a symbolic image.

A simple edge extraction method or a segmentation method could find the corner points (see Table 4-2). In this case, there is a direct match between the characteristics of the corners in the image and those of the object corners.

N	N	N	N	N	N	N	N	N	N	N	N	N
N	N	N	N	N	N	N	N	N	N	N	N	N
N	N	N	N	C	N	N	N	C	N	N	N	N
N	N	N	N	N	N	N	N	N	N	N	N	N
N	N	N	N	N	N	N	N	N	N	N	N	N
N	N	N	N	N	N	N	N	N	N	N	N	N
N	N	N	N	C	N	N	N	C	N	N	N	N
N	N	N	N	N	N	N	N	N	N	N	N	N
N	N	N	N	N	N	N	N	N	N	N	N	N

Table 4-2 Numeric digital intensity image of the corners of the image from Table 4-1 (C = corner; N = no corner).

However, a transformation process may be difficult. It may involve the recognition of a series of complex objects. Also, some objects could contain parts of other objects, there could be shadows, the light reflected by the object could vary, or the background could be restless.

The decision as to what type of transformation is suitable depends on the specific nature of the observation task, the image complexity, and the type of information available at the beginning.

In general, computer-supported object recognition and inspection is a complex task. It involves a number of different steps to successively transform object data into recognition information. To be suitable, a recognition method should include the following six steps: image formatting, conditioning, marking, grouping, extraction, and matching. These steps are shown schematically in Figure 4-13 and described briefly in the following section.

Computer-Assisted Graphics and Image Processing

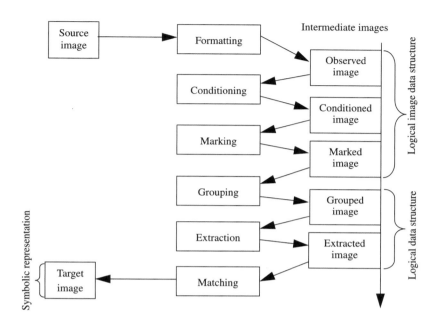

Figure 4-13 Steps involved in image recognition.

The Image-Recognition Procedure This section provides a brief introduction to the image recognition steps. A detailed analysis is given in [Nev82, HS92] and [GW93].

The formatting step shoots an image by use of a camera and transforms the image into digital form (i.e., pixels as described above). The conditioning, marking, grouping, extraction, and matching steps form a canonic division of the image recognition problem, where each step prepares and transforms the data required for the next step. Depending on the application, it may be necessary to apply this sequence to more than one level of recognition and description. These five steps are:

1. Conditioning: Conditioning is based on a model that assumes an image that can be observed is composed of information patterns, which are disturbed by irrelevant variations. Such variations are typically additive or multiplicative parts of such a pattern. Conditioning estimates the information pattern based on the observed image, so that noise (an unwanted disturbance that can be seen as a random change without recognizable pattern and that falsifies each measurement) can be suppressed. Also, conditioning can normalize the background by ignoring

irrelevant systematic or patterned variations. The typical conditioning process is independent from the context.

2. Marking: Marking is based on a model that assumes that information patterns have a structure in the form of spatial object arrangements, where each object is a set of interconnected pixels. Marking determines to which spatial objects each pixel belongs.

   An example of a marking operation is the edge recognition method described above. Edge recognition techniques find local discontinuities of some image attributes, such as intensity or color (e.g., the rim of a cup). These discontinuities are particularly interesting because there is a high probability that they occur along the boundaries of an object. Edge recognition finds a large quantity of edges, but not all of them are significant. For this reason, an additional marking operation has to be used after the edge recognition. This operation is called thresholding; it specifies the edges that should be accepted and those that can be discarded. This means that the operation filters and marks the significant edges of an image, while the remaining edges are removed. Additional marking operations can find corner dots.

3. Grouping: The grouping operation identifies objects marked in the previous step by grouping pixels that are part of the same object, or by identifying maximum quantities of interconnected pixels. Considering that intensity-based edge recognition is actually a gradual modification, the grouping step includes the edge connection step. A grouping operation that groups edges into lines is also called line fitting, for which the above described Hough transform and other transform methods can be used.

   The grouping operation changes the logical data structure. The original images, the conditioned, and marked images are all available as digital image data structures. Depending on the implementation, the grouping operation can produce either an image data structure in which an index is assigned to each pixel and the index is associated with this spatial occurrence, or a data structure that represents a quantity collection. Each quantity corresponds to a spatial occurrence and contains line/column pairs specifying the position, which are part of the occurrence. Both cases change the data structure. The relevant units are pixels before and pixel quantities after the grouping operation.

4. Extraction: The grouping operation defines a new set of units, but they are incomplete because they have an identity but no semantic meaning. The extraction operation calculates a list of properties for each pixel group. Such properties can include center (gravity), surface, orientation, spatial moments, spatial grayscale moments, and circles. Other properties depend on whether the group is seen as a region or as a circle sector. For a group in the form of a region, the number of holes in the interconnected pixel group could be a meaningful property. On the

other hand, the mean curvature could be a meaningful indicator for a circle section.

In addition, the extraction operation can measure topological or spatial relationships between two or more groups. For example, it could reveal that two groups touch each other, or that they are close spatial neighbors, or that one group has overlaid another one.

5. Matching: When the extraction operation is finished, the objects occurring in an image are identified and measured, but they have no content meaning. We obtain a content meaning by attempting an observation-specific organization in such a way that a unique set of spatial objects in the segmented image results in a unique image instance of a known object, for example, a chair or the letter "A". Once an object or a set of object parts has been recognized, we can measure things like the surface or distance between two object parts or the angles between two lines. These measurements can be adapted to a given tolerance, which may be the case in an inspection scenario. The matching operation determines how to interpret a set of related image objects, to which a given object of the three-dimensional world or a two-dimensional form is assigned.

While various matching operations are known, the classical method is template matching, which compares a pattern against stored models (templates) with known patterns and selects the best match.

4.3.2 Image Synthesis

Image synthesis is an integral part of all computer-supported user interfaces and a necessary process to visualize 2D, 3D, or higher-dimensional objects. A large number of disciplines, including education, science, medicine, construction, advertising, and the entertainment industry, rely heavily on graphical applications, for example:

- User interfaces: Applications based on the Microsoft Windows operating system have user interfaces to run several activities simultaneously and offer point-and-click options to select menu items, icons, and objects on the screen.
- Office automation and electronic publishing: The use of graphics in the production and distribution of information has increased dramatically since desktop publishing was introduced on personal computers. Both office automation and electronic publishing applications can produce printed and electronic documents containing text, tables, graphs, and other types of drawn or scanned graphic elements. Hypermedia systems allow viewing of networked multimedia documents and have experienced a particularly strong proliferation.
- Simulation and animation for scientific visualization and entertainment: Animated movies and presentations of temporally varying behavior of real and simulated objects on computers have been used increasingly for scientific visualization. For

example, they can be used to study mathematical models of phenomena, such as flow behavior of liquids, relativity theory, or nuclear and chemical reactions. Cartoon actors are increasingly modeled as three-dimensional computer-assisted descriptions. Computers can control their movements more easily compared to drawing the figures manually. The trend is towards flying logos or other visual tricks in TV commercials or special effects in movies.

Interactive computer graphics are the most important tool in the image production process since the invention of photography and television. With these tools, we cannot only create images of real-world objects, but also abstract, synthetic objects, such as images of mathematical four-dimensional surfaces.

4.3.2.1 Dynamic Versus Static Graphics

The use of graphics is not limited to static images. Images can also vary dynamically. For example, a user can control an animation by adjusting the speed or changing the visible part of a scene or a detail. This means that dynamics are an integral part of (dynamic) graphics. Most modern interactive graphics technologies include hardware and software allowing users to control and adapt the dynamics of a presentation:

- Movement dynamics: Movement dynamics means that objects can be moved or activated relative to a static observer's viewpoint. Objects may also be static while their environment moves. A typical example is a flight simulator containing components that support a cockpit and an indicator panel. The computer controls the movement of the platform, the orientation of the aircraft, and the simulated environment of both stationary and moving objects through which the pilot navigates.
- Adaptation dynamics: Adaptation dynamics means the current change of form, color, or other properties of observed objects. For example, a system could represent the structural deformation of an aircraft in the air as a response to many different control mechanisms manipulated by the user. The more subtle and uniform the change, the more realistic and meaningful is the result. Dynamic interactive graphics offer a wide range of user-controllable modes. At the same time, these modes encode and convey information, for example, 2D or 3D forms of objects in an image, their gray levels or colors, and the temporal changes of their properties.

4.4 Reconstructing Images

We said at the beginning of this chapter that a two-dimensional image signal taken by a camera is built by projecting the spatial (three-dimensional) real world onto the two-dimensional plane. This section introduces methods used to reconstruct the original three-dimensional world based on projection data. To reconstruct an image, we normally need a continuous series of projection images. The methods used to reconstruct

Reconstructing Images

images include the Radon transform and stereoscopy, which will be described briefly below.

4.4.1 The Radon Transform

Computer tomography uses pervious projections, regardless of whether or not an image is obtained from X-rays, ultrasound, magnetic resonance, or nuclear spin effects. The intensity of the captured image depends mainly on the "perviousness" of the volume exposed to radiation. A popular method used in this field is the Radon transform [Jäh97].

To understand how this method works, assume that the equation of a straight line across A and B runs at a distance d_1 and at an angle θ to the origin of $d_1 = r\cos\theta + s\sin\theta$.

We can form the integral of "brightness" along this beam by utilizing the fade-out property of the delta function, which is all zeros, except in the respective position, (r, s):

$$R_x(\theta, d_1) = \int_{-\infty}^{\infty}\int_{-\infty}^{\infty} x(r, s) \cdot \delta(r\cos\theta + s\sin\theta - d_1)drds; \quad (0 \leq \theta \leq \pi)$$

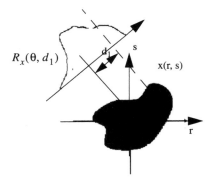

Figure 4-14 Using the Radon transform.

If we operate this transform on a variable instead of a fixed distance, d_1, we obtain the continuous two-dimensional Radon transform $x(\theta, d)$ from $x(r, s)$. Two variants of the Radon transform are in use: parallel-beam projection and fan-beam projection $x(\theta, t)$, which is based on a punctual radiation source, where angle β has to be defined for each single beam.

In addition, the method shows that the 1D Fourier transform of a Radon transform is directly related to the 2D Fourier transform of the original image in all possible angles θ (this is the projection disk theorem):

$$F(u, v) = F\{x(r, s)\}; \quad F(\theta, w) = F\{R_x(\theta, t)\}$$
$$F(u, v) = F(\theta, w) \quad \text{for} \quad u = w\cos\theta, \; v = w\sin\theta$$

We can now clearly reconstruct the image signal from the projection:

$$x(r, s) = \int_0^{2\pi} \int_0^{\infty} F(\theta, w) e^{j2\pi w(r\cos\theta + s\sin\theta)} w \, dw \, d\theta$$

$$= \int_0^{\pi} \int_{-\infty}^{\infty} F(\theta, w)|w| e^{j2\pi w d} \, dw \, d\theta$$

However, reconstruction will be perfect only for the continuous case. In the discrete case, the integrals flow into sums. Also, only a limited number of projection angles is normally available, and inter-pixel interpolations are necessary. As a general rule, reconstruction improves (but gets more complex) as more information becomes available and the finer the sampling in d and θ (and the more beams available from different directions).

4.4.2 Stereoscopy

Objects in natural scenes observed by humans are normally not transparent; instead, we see only their surfaces. The Radon transform does not make sense in this case, because only the brightness of a single point will flow into the projection. The principle of stereoscopy, which is necessary in order to be able to determine the spatial distance, is based on the capture of two "retinal beams." This process typically involves a central projection equation and two cameras. Assume that one camera is positioned at $W_1=0$, $W_2=0$, $W_3=0$ and the second at $W_1=A$, $W_2=B$, $W_3=0$. Further assume that a point, P, of the world coordinate $[W_1, W_2, W_3]$ is in the image level of the first camera at position $[r_1, s_1]$ and for the second camera at position $[r_2, s_2]$. Using the central projection equation, we obtain

$$W_1 = r_1 \cdot \frac{W_3}{F_1} = A + r_2 \cdot \frac{W_3}{F_2}; \quad W_2 = s_1 \cdot \frac{W_3}{F_1} = B + s_2 + \frac{W_3}{F_2}$$

Assuming that both cameras have the same focal length, F, we obtain the distance of point P in W_3:

$$W_3 = F \cdot \frac{A}{(r_1 - r_2)} = F \cdot \frac{B}{(s_1 - s_2)}$$

The relative shift between the observation points on the image levels of a camera is called stereoscopic parallax:

$$r_1 - r_2 = F \cdot \frac{A}{W_3}; \quad s_1 - s_2 = F \cdot \frac{B}{W_3}$$

The stereoscopic parallax is proportionally inverse to the distance of the point observed by the cameras. We now obtain

$$W_1 = r_1 \cdot \frac{A}{r_1 - r_2}; \quad W_2 = s_1 \cdot \frac{B}{s_1 - s_2}$$

This example shows that it is not difficult to produce an exact image of the three-dimensional space. However, the process is limited, for example with regard to the estimated height (W_2), when B is very small (both eyes or cameras are at same height). Another problem is that we always have to produce an exact point-by-point correspondence between the image dots captured by the cameras, which is not possible along the object boundaries, where the background may be obstructed for one camera but not for the other one (see Figure 4-15). Yet another problem is that camera and reflection noise may cause differences in the local brightness.

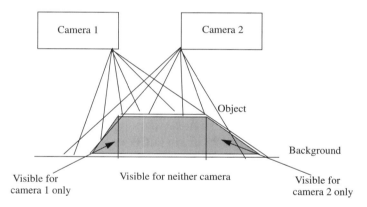

Figure 4-15 Stereovision.

4.5 Graphics and Image Output Options

Modern visual output technologies use raster display devices that display primitives in special buffers called refresh buffers in the form of pixel components. Figure 4-16 shows the architecture of a raster display device.

Some raster display devices have a display controller in hardware that receives and interprets output command sequences. In low-end systems, for example, personal computers (as in Figure 4-16), the display controller is implemented in software and exists as a component of a graphics library. In this case, the refresh buffer is part of the CPU memory. The video controller can read this memory to output images.

The complete display of a raster display device is formed by the raster, consisting of horizontal raster lines that are each composed of pixels. The raster is stored in the form of a pixel matrix that represents the entire screen area. The video controller samples the entire screen line by line.

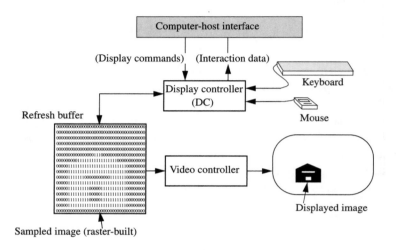

Figure 4-16 Architecture of a raster display device.

Raster graphics can represent areas filled with colors or patterns, for example, realistic images of three-dimensional objects. The screen refresh process is totally independent of the image complexity (e.g., number of polygons), because modern hardware works fast enough to read each pixel from the buffer in every refresh cycle.

4.5.1 Dithering

Raster graphics technology has been developed at a fast pace, so that all modern computer graphics can represent color and gray-level images. The color depends primarily on the object, but also on light sources, environment colors, and human perception. On a black-and-white TV set or computer monitor, we perceive achromatic light, which is determined by the light quality. Light quality is an attribute determined by brightness and luminance parameters.

State-of-the-art technology uses the capabilities of the human eye to provide spatial integration. For example, if we perceive a very small area at a sufficiently long distance, our eyes form a mean value of fine-granular details of that small area and acquire only its overall intensity. This phenomenon is called continuous-tone approximation (dithering by means of grouped dots). This technique draws a black circle around each smallest resolution unit, where the circle's surface is proportional to the blackness, *1-I* (*I* stands for intensity), of the original picture surface. Image and graphics output devices can approximate variable surface circles for continuous-tone reproduction. To understand how this works, imagine a 2×2-pixel surface displayed on a monitor that supports only two colors to create five different gray levels at the cost of splitting the spatial resolution along the axes into two halves. The patterns shown in Figure 4-17 can be filled with 2×2 surfaces, where the number of enabled pixels is proportional to the desired intensity. The patterns can be represented by a dithering matrix.

Figure 4-17 Using 2∞×2 dithering patterns to approximate five intensity levels.

This technique is used by devices that cannot represent individual dots (e.g., laser printers). Such devices produce poor quality if they need to reproduce enabled isolated pixels (the black dots in Figure 4-17). All pixels enabled for a specific intensity must be neighbors of other enabled pixels.

In contrast, a CRT display unit is capable of representing individual dots. To represent images on a CRT display unit, the grouping requirement is less strict, which means that any dithering method based on a finely distributed dot order can be used. Alternatively, monochrome dithering techniques can be used to increase the number of available colors at the cost of resolution. For example, we could use 2×2 pattern surfaces to represent 125 colors on a conventional color screen that uses three bits per pixel—one each for red, green, and blue. Each pattern can represent five intensities for each color by using the continuous-tone pattern shown in Figure 4-17, so that we obtain 5×5×5 = 125 color combinations.

4.6 Summary and Outlook

This chapter described a few characteristics of images and graphics. The quality of these media depends on the quality of the underlying hardware, such as digitization equipment, monitor, and other input and output devices. The development of input and output devices progresses quickly. Most recent innovations include new multimedia devices and improved versions of existing multimedia equipment. For example, the most recent generation of scanners for photographs supports high-quality digital images. Of course, such devices are quickly embraced by multimedia applications. On the other hand, the introduction of new multimedia devices (e.g., scanners) implies new multimedia formats to ensure that a new medium (e.g., photographs) can be combined with other media. One example is Kodak's introduction of their Photo Image Pac file format. This disk format combines high-resolution images with text, graphics, and sound. It allows the development of interactive photo-CD-based presentations [Ann94b]. (Optical storage media are covered in Chapter 8.)

Most existing multimedia devices are continually improved and new versions are introduced to the market in regular intervals. For example, new 3D digitization cards allow users to scan 3D objects in any form and size into the computer [Ann94a].

CHAPTER 5

Video Technology

In addition to audio technology, television and video technology form the basis of the processing of continuous data in multimedia systems. In this chapter, we consider concepts and developments from this area that are significant for a basic understanding of the video medium. For further details, we make reference to standard works in the area of television technology (see [Joh92]).

Video data can be generated in two different ways: by recording the real world and through synthesis based on a description. We begin our discussion of video technology by covering current and future video standards (analog and digital) with respect to properties of human perception.

5.1 Basics

The human eye is the human receptor for taking in still pictures and motion pictures. Its inherent properties determine, in conjunction with neuronal processing, some of the basic requirements underlying video systems.

5.1.1 Representation of Video Signals

In conventional black-and-white television sets, the video signal is usually generated by means of a Cathode Ray Tube (CRT).

In order to lay the groundwork for a later understanding of the transmission rates of films, we cover television signals in detail here, although we do not consider camera or monitor technology. We begin by analyzing the video signal produced by a camera and the resulting pictures [BF91].

The representation of a video signal comprises three aspects: visual representation, transmission, and digitization.

5.1.1.1 Visual Representation

A key goal is to present the observer with as realistic as possible a representation of a scene. In order to achieve this goal, the television picture has to accurately convey the spatial and temporal content of the scene. Important measures for this are:

- Vertical details and viewing distance

The geometry of a television image is based on the ratio of the picture width W to the picture height H. This width-to-height ratio is also called the aspect ratio. The conventional aspect ratio (for television) is 4/3=1.33. Figure 5-1 shows an example of this ratio.

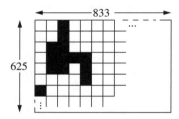

Figure 5-1 Decomposition of a motion picture. Width to height in ratio 4:3.

The viewing distance D determines the angular field of view. This angle is usually calculated as the ratio of the viewing distance to the picture height (D/H).

The smallest detail that can be reproduced in a picture is a pixel (picture element). Ideally every detail of a picture would be reproduced by a picture element. In practice, however, it is unavoidable that some details lie between scan lines. For such picture elements, two scan lines are necessary. The result is loss of vertical resolution. Measurements of this effect show that only about 70 percent of the vertical details are present in the scan lines. This ratio, known as the Kell factor, is independent of the sampling method, that is, whether the scan lines follow one another sequentially (progressive sampling) or alternately (interlaced sampling).

- Horizontal detail and picture width

The picture width normally used for television is 4/3 times the picture height. The horizontal field of view can be determined using the aspect ratio.

- Total detail content of a picture

The vertical resolution is equal to the number of picture elements of the picture height, while the number of horizontal picture elements is equal to the product of the vertical resolution and the aspect ratio. The product of the picture's elements vertically and horizontally is the total number of picture elements in the image. However, in the case of television pictures, not all lines (and columns) are visible to the observer. The invisible areas are often used to transmit additional information.

- Depth perception

In nature, humans perceive the third dimension, depth, by comparing the images perceived by each eye, which view from different angles. In a flat television picture, a considerable portion of depth perception is derived from the perspective appearance of the subject matter. Further, the choice of the focal length of the camera lens and changes in depth of focus influence depth perception.

- Luminance

Color perception is achieved by three signals, proportional to the relative intensities of red, green, and blue light (RGB) present in each portion of the scene. These are conveyed to the monitor separately and the tube reproduces them at each point in time (unlike a camera). Often a different signal division is used for transmission and storage: one brightness signal (luminance) and two color difference signals (chrominance). This division will be explained in more detail below.

- Temporal aspects of illumination

Another property of human visual perception is the limit of motion resolution. In contrast to the continuous pressure waves of an acoustic signal, a discrete sequence of individual still pictures is perceived as a continuous sequence. This property is used in television, in films, and for video data in computer systems. The impression of motion is created by presenting a rapid succession of barely differing still pictures (frames). Between frames, the light is cut off briefly. Two conditions must be met in order to represent a visual reality through motion pictures. First, the rate of repetition of the images must be high enough to ensure continuity of movements (smooth transition) from frame to frame. Second, the rate must be high enough that the continuity of perception is not disrupted by the dark intervals between pictures.

- Continuity of motion

It is known that continuous motion is only perceived as such if the frame rate is higher than 15 frames per second. To make motion appear smooth, at least 30 frames per second must be used if the scene is filmed by a camera and not generated synthetically. Films recorded using only 24 frames per second often appear strange, especially when large objects move quickly and close to the viewer, as in pan shots. Showscan [Dep89] is a technology for producing and presenting films at 60 frames per second using 70-millimeter film. This scheme produces a large image that occupies a greater portion of the field of view, resulting in smoother motion.

There are various standards for motion video signals that establish frame rates ensuring suitable continuity of motion. The standard used in the United States, NTSC (National Television Systems Committee), originally set the frame rate at 30 Hz. This was later changed to 29.97 Hz in order to fix the separation between the visual and audio carriers at precisely 4.5 MHz. NTSC scanning equipment represents frames using the 24 Hz standard by translating them to the 29.97 Hz scanning rate. A European standard for motion video, PAL (Phase Alternating Line), adopted a repetition rate of 25 Hz, but uses a frame rate of 25 Hz.

- Flicker

If the refresh rate is too low, a periodic fluctuation of the perceived brightness can result. This is called the flicker effect. The minimum refresh rate to avoid flicker is 50 Hz. Achieving continuous, flicker-free motion would thus require a high refresh rate. However, in both movies and television, there are technical measures that allow lower refresh rates to be used.

The flicker effect would be very disturbing in films with, for example, 16 pictures per second without any additional technical measures. In order to reduce flicker, the light is interrupted an additional two times during the projection of a frame, yielding a picture refresh rate of $3 \times 16\,\text{Hz} = 48\,\text{Hz}$.

In television, display refresh buffers—expensive until recently—can be used to alleviate the flicker effect. Picture data are written into the buffer at a rate higher than needed for motion resolution (e.g., 25 Hz). The monitor reads the display data out at a rate that eliminates the flicker effect (e.g., 70 Hz). This corresponds to the 70 Hz refresh rate of higher quality computer screens.

In television, the full picture is divided into two half pictures consisting of interleaved scanning lines. One half of the picture is transmitted after the other using interlaced scanning. The transmission of full pictures takes place at around 30 Hz (exactly 29.97 Hz), or 25 Hz in Europe, whereas the transmission of half pictures takes place at $2 \times 30\,\text{Hz} = 60\,\text{Hz}$ or $2 \times 25\,\text{Hz} = 50\,\text{Hz}$, respectively. Figure 5-2 shows an example of this. Visual perception drops considerably more with a refresh rate of 25 Hz (unbroken line) than 50 Hz.

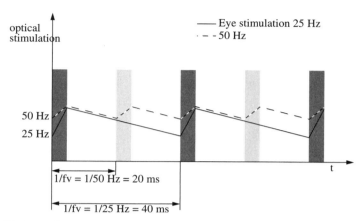

Figure 5-2 Flicker effect. Eye stimulation with refresh rates of 25 Hz and 50 Hz.

5.1.2 Signal Formats

Video signals are often transmitted to the receiver over a single television channel. In order to encode color, consider the decomposition of a video signal into three subsignals. For reasons of transmission, a video signal is comprised of a luminance signal and two chrominance (color) signals. In NTSC and PAL systems, the component transfer of luminance and chrominance in a single channel is accomplished by specifying the chrominance carrier to be an odd multiple of half the line-scanning frequency. This causes the component frequencies of chrominance to be interleaved with those of luminance. The goal is to separate the sets of components in the receiver and avoid interference between them before the primary color signals are recovered for display. In practice, however, there are degradations in the picture quality, known as color crosstalk and luminance crosstalk. These effects have led the manufacturers of NTSC receivers to reduce the luminance bandwidth to less than 3 MHz, under the carrier frequency of 3.58 MHz and far below the broadcast signal theoretical maximum limit of 4.2 MHz. This limits the vertical resolution in such devices to about 25 lines. Chrominance and luminance signals are separated by using a simple notch filter tuned to the subcarrier's frequency. Today comb filters are also used for this purpose. The transmitter also uses a comb filter in the coding process.

Several approaches to color encoding are described below.

5.1.2.1 Color Encoding

- RGB signal

  An RGB signal consists of separate signals for red, green, and blue. Every color can be encoded as a combination of these three primary colors using additive color mixing. The values R (for red), G (for green), and B (for blue), are normalized such that white results when $R + G + B = 1$ in the normalized representation.

- YUV signal

Since human vision is more sensitive to brightness than to color, a more suitable encoding separates the luminance from the chrominance (color information). Instead of separating colors, the brightness information (luminance Y) is separated from the color information (two chrominance channels U and V). For reasons of compatibility with black-and-white receivers, the luminance must always be transmitted. For black-and-white reception, the utilization of the chrominance components depends on the color capabilities of the television set.

The YUV signal can be calculated as follows:

$Y = 0.30 R + 0.59 G + 0.11 B$

$U = (B - Y) \times 0.493$

$V = (R - Y) \times 0.877$

An error in the resolution of the luminance (Y) is more serious than one in the chrominance values (U, V). Thus the luminance values can be encoded using higher bandwidth than is used for the chrominance values.

Due to the different component bandwidths, the encoding is often characterized by the ratio between the luminance component and the two chrominance components. For example, the YUV encoding can be specified as a $(4:2:2)$ signal. Further, the YUV encoding is sometimes called the Y, $B-Y$, $R-Y$ signal, from the dependencies among U, $B-Y$, V and $R-Y$ in the equations above.

- YIQ signal

A similar encoding exists for NTSC's YIQ signal:

$Y = 0.30 R + 0.59 G + 0.11 B$

$I = 0.60 R - 0.28 G - 0.32 B$

$Q = 0.21 R - 0.52 G + 0.31 B$

A typical NTSC encoder is shown in Figure 5-3. It produces the I and Q signals, then performs a quadrature amplitude modulation on the suppressed chrominance subcarrier and adds the modulated signal to the luminance Y. The signals are also blanked and synchronized.

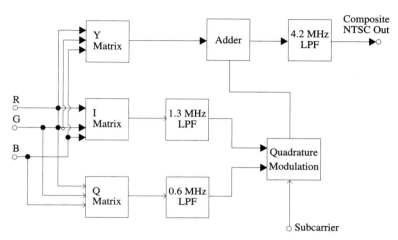

Figure 5-3 YIQ encoding operations of an NTSC system.

5.1.2.2 Composite Signal

An alternative to component encoding is to combine all information in one signal. This implies that the individual components (RGB, YUV, or YIQ) must be combined in a single signal. The basic information consists of luminance information and chrominance difference signals. However, the luminance and chrominance signals can interfere since they are combined into one signal. For this reason, television technology uses appropriate modulation methods aimed at eliminating this interference.

The basic bandwidth needed to transmit the luminance and chrominance signals for the NTSC standard is 4.2 MHz.

5.1.2.3 Computer Video Format

The video format processed by a computer depends on the video input and output devices. Current video digitalization hardware differs with respect to the resolution of the digital images (frames), quantization, and the frame rate (frames/second). Motion video output depends on the display hardware used, usually a raster display. The typical architecture of such a device is shown in Figure 5-4.

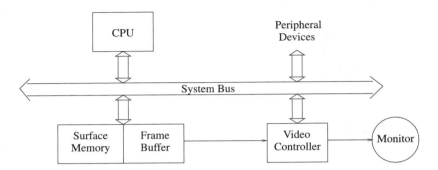

Figure 5-4 Architecture of a raster display.

The video controller displays the image stored in the frame buffer, accessing the buffer through a separate port as often as required by the video scanning rate. The most important task is the constant refresh of the display. Due to the disturbing flicker effect, the video controller cycles through the frame buffer, one scan line at a time, typically 60 times/second. To display different colors on the screen, the system works with a Color Look-Up Table (CLUT or LUT). At any given time, a limited number of colors (n) are available for the whole picture. The set of the n most frequently used colors is chosen from a color palette consisting of m colors, whereby in general $n \ll m$.

Some examples of well-known computer video formats are presented here. Each system supports various resolutions and color presentations.

- The Color Graphics Adapter (CGA) has a resolution of 320×200 pixels with simultaneous display of four colors, among other modes. The necessary storage capacity per frame is thus

$$320 \times 200 \text{ pixel} \times \frac{2 \text{ bit/pixel}}{8 \text{ bit/byte}} = 16{,}000 \text{ bytes}$$

- The Enhanced Graphics Adapter (EGA) supports display resolution of 640×350 pixels with 16 simultaneous colors. The necessary storage capacity per frame is

$$640 \times 350 \text{ pixel} \times \frac{4 \text{ bit/pixel}}{8 \text{ bit/byte}} = 112{,}000 \text{ bytes}$$

- The Video Graphics Array (VGA) works mostly with a resolution of 640×480 pixels with 256 simultaneous colors. The monitor is controlled via an analog RGB output. The necessary storage capacity per frame is

$$640 \times 480 \text{ pixel} \times \frac{8 \text{ bit/pixel}}{8 \text{ bit/byte}} = 307{,}200 \text{ bytes}$$

- The Super Video Graphics Array (SVGA) can present 256 colors at a resolution of 1,024×768 pixels. The necessary storage capacity per frame is

$$1{,}024 \times 768 \text{ pixel} \times \frac{8 \text{ bit/pixel}}{8 \text{ bit/byte}} = 786{,}432 \text{ bytes}$$

Other SVGA modes include 1,280×1,024 pixels and 1,600×1,280 pixels.

SVGA video adapters are available with video accelerator chips that overcome reduced performance at higher resolution and/or higher numbers of colors [Lut94]. Video accelerator chips can be used to improve playback of video, which would normally appear in a window of at most 160×120 pixels. A video accelerator chip allows for playback of recorded video sequences at a significantly higher rate and quality [Ann94c].

5.2 Television Systems

Television is one of the most important applications driving the development of motion video. Since 1953, television has undergone many far-reaching changes. This section provides an overview of television systems, encompassing conventional black-and-white and color systems, enhanced resolution television systems intended as an intermediate solution, and digital interactive video systems and Digital Video Broadcasting (DVB).

5.2.1 Conventional Systems

Black-and-white and current color television is based on the properties of video signals as described in Section 5.1.1. Early on, different parts of the world adopted different video standards. Conventional television systems use the following standards:

- NTSC stands for National Television Systems Committee and is the oldest and most widely used television standard. The standard originated in the US and uses color carriers of approximately 4.429 MHz or approximately 3.57 MHz. NTSC uses quadrature amplitude modulation with a suppressed color carrier and a refresh rate of about 30 Hz. A picture consists of 525 rows.

  NTSC can use 4.2 MHz for the luminance and 1.5 MHz for each of the two chrominance channels. Television sets and video recorders use only 0.5 MHz for the chrominance channels.

- SECAM stands for Sequential Couleur avec Memoire and is used primarily in France and Eastern Europe. In contrast to NTSC and PAL, it is based on frequency modulation. Like PAL, SECAM uses a refresh rate of 25 Hz. Each picture consists of 625 rows.

- PAL stands for Phase Alternating Line and was proposed in 1963 by W. Bruch of Telefunken. It is used in parts of Western Europe. Figure 5-5 is a schematic of the picture preparation process.

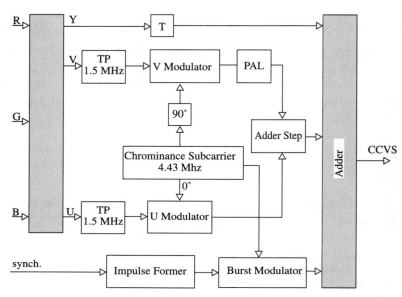

Figure 5-5 Color signal preparation in the PAL standard: converting RGB to CCVS.

The color carrier lies in the CCVS (Color Composite Video Signal) spectrum about 4.43 MHz from the picture carrier. The basic principle is quadrature amplitude modulation, whereby the color carrier and the chrominance signal U are directly multiplied. The color carrier, shifted by 90 degrees, is then multiplied by the chrominance signal V. This product is then added to the shifted color carrier. This is a normal quadrature amplitude modulation. In every other row, in addition to the actual quadrature amplitude modulation, the phase of the modulated V signal is rotated in order to reduce phase errors.

5.2.2 High-Definition Television (HDTV)

Research in the area of High-Definition Television (HDTV) began in Japan in 1968. This phase is considered to be the third technological change in television, after black-and-white and the introduction of color television. HDTV strives for picture quality at least as good as that of 35 mm film.

Promoters of HDTV pursued the goal of approaching integrating the viewer with the events taking place on the screen [Hat82]. Television systems, filming techniques, and viewing requirements were chosen in order to give the viewer the impression of being involved in the scene.

The parameters that had to be defined in order to achieve this goal were resolution, frame rate, aspect ratio, interlaced and/or progressive scanning formats, and viewing conditions [Org96].

- *Resolution*

  Compared to conventional systems, an HDTV picture has about twice as many horizontal and vertical columns and lines, respectively. The improved vertical resolution is achieved by using more than 1,000 scanning lines. Improved luminance details in the picture can be accomplished with a higher video bandwidth, about five times that used in conventional systems. Two resolution schemes are recommended for practical applications: the so-called "High 1440 Level" with $1,440 \times 1,152$ pixels and the "High Level" containing $1,920 \times 1,152$ pixels.

- *Frame rate*

  The number of frames per second was bitterly discussed in the ITU Working Groups. For practical reasons, namely compatibility with existing TV systems and with movies, agreement on a single HDTV standard valid worldwide could not be achieved; options of 50 or 60 frames per second were established. Newly developed, very efficient standard-conversion techniques, based in part on movement estimation and compensation, can mitigate this problem.

- *Aspect ratio*

  The aspect ratio is defined as the ratio of picture width to picture height. Originally, a ratio of $16:9 = 1.777$ was adopted; the ratio in current televisions is $4:3$.

- *Interlaced and/or progressive scanning formats*

  Conventional TV systems are based on alternation of scanning lines—each frame is composed of two consecutive fields, each containing half the scanning lines of a picture, which are scanned and presented in interlaced mode. In progressive scanning, used for example in computer displays, there is only one such field per picture and the number of scanning lines per field is doubled.

- *Viewing conditions*

  The field of view and thus the screen size play an important role in visual effects and thus also for the feeling of "reality." Early studies found that the screen area must be bigger than $8,000 \text{cm}^2$. The line count per picture is about twice as large as in conventional television, so the normal viewing distance can be halved compared to conventional systems and results in three times the picture height ($3H$).

In 1991, researchers at NTT Labs reported further progress in HDTV technology [OO91]. Their concept integrated various video media with their corresponding quality levels (communication, broadcast, and display) into one system. In order to achieve this integration, a minimum spatial resolution of $2k \times 2k$ pixels is needed, along with temporal resolution of at least 60 frames per second and a signal resolution of at least 256 steps (corresponding to 8 bits). A resolution of 2k×2k corresponds to a high-resolution photo or a color A4 print.

System	Total channel width (MHz)	Video basebands (MHz)			Frame rate (Hz)	
		Y	R - Y	B - Y	Recording: camera	Playback: monitor
HDTV (USA)	9.0	10.0	5.0	5.0	59.94-p	59.94-i
NTSC	6.0	4.2	1.0	0.6	59.94-i	59.94-i
PAL	8.0	5.5	1.8	1.8	50-i	50-i
SECAM	8.0	6.0	2.0	2.0	50-i	50-i

Table 5-1 Properties of TV systems (p: progressive, i: interlaced).

5.3 Digitization of Video Signals

Before a motion picture can be processed by a computer or transmitted over a network, it must be converted from an analog to a digital representation.

This digitization process consists of the three steps of sampling, quantization, and coding. In determining the sampling frequency, the Nyquist Theorem must be followed. This theorem states that the signal being sampled cannot contain any frequency components that exceed half the sampling frequency. In order to prevent the baseband from overlapping with repeating spectra and to allow for real hardware components not behaving ideally, the sampling rate is normally chosen somewhat higher than the limit dictated by the Nyquist Theorem.

Since the gray value of a sampled spot can take on any value in a continuous range, it must be quantized in order to be processed digitally. The gray-scale is subdivided into several ranges, and each pixel is assigned only one of these values. In order to achieve acceptable quality in an image reconstructed from quantized pixels, it is a good idea to use at least 256 quantization steps.

The CCIR (Consultative Committee International Radio, currently ITU) had already passed an international standard for digital television in 1982. This standard (ITU 601) describes specific scanning resolutions and covers scanning and coding. There are two possible types of digital coding: composite coding and component coding.

5.3.1 Composite Coding

The simplest way to digitize a video signal is to sample the entire analog signal (CCVS—Color Composite Video Signal). Here all signal components are converted together to a digital representation. This "integrated coding" of an entire video signal is fundamentally simpler than digitizing separate signal components (luminance signal and two chrominance signals). However this approach also has numerous drawbacks:

- There is frequently disturbing crosstalk between the luminance and chrominance information.
- The composite coding of a television signal depends on the television standard used. This would lead to a further difference among the various standards, in addition to different numbers of scanning lines and motion frequencies. Even if multiplexing methods were used for further signal transmission, the standards difference would be bothersome since different transmission techniques would have to be used for different digital television standards.
- Because the luminance information is more important than the chrominance information, it should also take more bandwidth. When composite coding is used, the sampling frequency cannot be adapted to bandwidth requirements of different components. If component coding is used, it is possible to decouple the sampling frequency from the color carrier frequency.

Having detailed the disadvantages of composite coding, we now consider the properties of component coding.

5.3.2 Component Coding

Component coding is based on the principle of separate digitization of the components, that is, the luminance and color difference signals. These can then be transmitted together using a multiplexing method. The luminance signal (Y), which is more important than the chrominance signal, is sampled at 13.5 MHz, while the chrominance signals ($R-Y$, $B-Y$) are sampled at 6.75 MHz. The digitized luminance and chrominance signals are uniformly quantized using 8 bits. Due to the different component bandwidths in the ratio 4:2:2, this technique produces 864 sampling values per line for luminance (720 are visible) and 432 for each of the chrominance components (360 are visible). In PAL, for example, a frame consists of 575 lines and there are 25 full frames per second. The high data rate and the fact that these signals do not fit in the PCM hierarchy (139.264 Mbit/s, 34.368 Mbit/s) are problematic. Thus several substandards with lower data rates have been defined. The data rates are easy to derive from the bandwidths of the components (13.5 MHz, 6.75 MHz, 6.75 MHz).

The standardized sampling frequency of 13.5 MHz for R, G, B, or Y baseband signals [CCI82] and the 8-bit quantization result in a data rate of 108 Mbit/s for each individual signal. For the chrominance components Cr and Cb, the reduced sampling rate of 6.75 MHz corresponds to 54 Mbit/s. The sampling frequency of 13.5 MHz is defined as an integral multiple in both the 625- and the 525-line TV standards.

Signals	Sampling frequency [MHz]	Samples/line	Lines	Data rate [Mbit/s]	Total rate [Mbit/s]	Format
R	13.5	864	625	108		4:4:4
G	13.5	864	625	108		ITU 601
B	13.5	864	625	108	324	
Y	13.5	864	625	108		4:2:2
Cr	6.75	432	625	54		ITU 601
Cb	6.75	432	625	54	216	
Y	13.5	720	576	83		4:2:2
Cr	6.75	360	576	41.5		
Cb	6.75	360	576	41.5	166	
Y	13.5	720	576	83		4:2:0
Cr/	6.75	360	576	41.5		
Cb					124.5	
Y	6.75	360	288	20.7		4:2:0
Cr/	3.375	180	288	10.4		SIF
Cb					31.1	

Table 5-2 Comparison of component coding.

The resulting data rates for transmitting *R, G, B* (format 4:4:4) and *Y, Cr, Cb* (format 4:2:2), are 324 Mbit/s and 216 Mbit/s, respectively. These relationships are listed in Table 5-2. In the 601 standard there is an option of using 10-bit signal resolution for broadcast equipment and studio connections; the data rate increases accordingly. For the 16:9 aspect ratio, a sampling rate of 18 MHz instead of 13.5 MHz was proposed in order to provide sufficient bandwidth for this new picture format [CCI82]. This increases the data rate by a factor of 4/3.

Assuming a simple data reduction scheme is used, for example skipping the blanking intervals, the total data rate can be reduced from 216 Mbit/s to 166 Mbit/s, and further to 124.5 Mbit/s using line-sequential *Cr/Cb* transmission (format 4:2:0). Reducing sampling (subsampling) in the horizontal and vertical directions by a factor of two yields a data rate of 31.1 Mbit/s (so-called Source Input Format, SIF). SIF is a special picture format with progressive line scanning (no interlace) defined to work independently of the presentation. All of the values listed in Table 5-2 assume a frame rate of 25 Hz and 8-bit quantization.

5.4 Digital Television

When the topic of digital television was taken up by the relevant Working Groups of the ITU, discussion centered around the digital representation of television signals, that is, composite coding or component coding. After considerable progress in compression technology (see Chapter 7) and an agreement to use exclusively component coding, at least in the TV studio, the work concentrated on the distribution of digital signals and consequently took on more of a system-oriented viewpoint.

In Europe, the development of digital television (Digital TeleVision Broadcasting, DTVB or—less precisely—Digital Video Broadcasting, DVB) started in the early 90s. At this time, universities, research facilities, and businesses were already intensely working on pushing a European HDTV system. Interested partners founded a European consortium—the European DVB Project—that made quick headway upon beginning work and thus played a major role in the preliminary standardization work of all digital television system components. Excluded, however, are studio and display technologies [FKT96].

One of the first important decisions was the selection of MPEG-2 for the source coding of audio and video data and of the MPEG-2 system technology for the creation of elementary program streams and transport streams (see Compression in Section 7.7). Although the original MPEG-2 standard [Org96] met practical requirements, it was too broad to be implemented economically in its entirety. As a result, the syntax and possible parameters were restricted; these DVB recommendations are contained in the "Guidelines Document" [Ins94].

Typical documents that describe the DVB system components are [Ins95] and [Eur94]. The former, also called DVB-TXT, specifies how to handle "analog" teletext in a DVB environment. A mechanism for transmitting all types of subtitles and graphical elements as part of the DVB signal is described in [Eur96]. Program descriptions and navigation tools are covered by the Service Information Document (SI) [Eur94].

Satellite connections, CATV networks, and (S)MATV ((Small) Master Antenna TV) systems are best suited for distributing digital television signals. Terrestrial broadcast services, for example, data distribution to households over telephone connections or using "Multichannel Microwave Distribution Systems" (MMDS), are other technical possibilities. Suitable transmission systems had to be developed for all these options and standardized by the DVB Project. The standard for terrestrial broadcast (DVB-T) is [Eur96].

The European Telecommunications Standards Institute (ETSI) adopted the DVB system digital broadcast proposals for satellites (DVB-S) and for CATV systems (DVB-C) as official standards. These are ETS 300421 and ETS 300429. The standards documents ETS 300472, ETS 300468, and ETS 300473 apply to (S)MATV.

If microwaves are used to transmit DVB signals, there are two specifications, depending on the frequency range used. ETSI standards also describe MMDS for use at frequencies above 10 GHz (DVB-MS). This transmission system is based on the use of DVB-S technology. ETS 749 is applicable to frequencies below 10 GHz. This specification is based on DVB-C technology and is thus also designated as DVB-MC.

Additionally, recommendations were published for conditional access, for establishing backward channels in interactive video applications, and for the private use of specific networks.

In conclusion, the DVB solutions for digital television afford many advantages, the most important being the following:

- the increased number of programs that can be transmitted over a television channel,
- the option of adapting video and audio quality to each application,
- the availability of exceptionally secure encryption systems for pay-TV services,
- the availability of tools to develop and implement new services such as data broadcast, multimedia broadcast, and video-on-demand, and
- the option of integrating new Internet services representing the "convergence" of computers and TV.

CHAPTER 6

Computer-Based Animation

To animate something is, literally, to bring it to life. An animation covers all changes that have a visual effect. Visual effects can be very different attributes: positions (motion dynamics), form, color, transparency, structure, and texture of an object (update dynamics), as well as changes in lighting, camera position, orientation, and focus.

This chapter covers primarily computer-based animation, since this type of animation constitutes a "medium" of integrated multimedia systems. Today, computer-based animations are produced, edited, and generated with the help of a computer using graphical tools to create visual effects. Naturally, the discipline of traditional, noncomputer-based animation continues to exist. Interestingly, many steps of conventional animation appear to be ideally suited for support by computer.

6.1 Basic Concepts

6.1.1 Input Process

Before the computer can be used, drawings must be digitized to create key frames, where the entities being animated are at extreme or characteristic positions. These digitized images can be produced by the computer using appropriate programs or created by digitizing photos (pictures of real objects) or drawings. The drawings may need to be carefully post-processed (e.g., filtering) in order to clean up any glitches arising from the input process.

6.1.2 Composition Stage

Individual frames in a completed animation are generated by using image composition techniques to combine foreground and background elements [FDFH92]. By placing low-resolution digitized frames of an animation in a grid, a trailer film (pencil test) can be generated using the pan-zoom feature available in some frame buffers. The frame buffer can take a particular portion of such an image (pan) and enlarge it up to the size of the whole screen (zoom). This process can be repeated with the various elements contained in an animation's frames. If this procedure is done fast enough, then the effect of continuity results. Since each frame of an animation is reduced to a small portion of the total image (1/25 or 1/36) and then enlarged to the full image size, the resolution of the monitor can be effectively reduced.

6.1.3 Inbetween Process

The animation of movement from one position to another requires the composition of frames with intermediate positions (intermediate frames) between key frames. In computer-based animation, this inbetween processing is done using interpolation methods. In interpolation, the system obtains only the beginning and end positions. Linear interpolation, sometimes called *lerping,* is the simplest method, but is subject to numerous limitations. For example, if one uses lerping to calculate the intermediate positions of a ball that has been thrown in the air and uses only three key frames, as shown in Figure 6-1 (a), then the resulting motion of the ball, depicted in Figure 6-1 (b), is totally unrealistic.

Figure 6-1 Linear interpolation of the motion of a ball: (a) key frames, (b) additional intermediate frames.

Due to the disadvantages of lerping, splines are often used to smooth out the interpolation between key frames. Splines can be used to smoothly vary different parameters as a function of time. With splines, individual points (or individual objects) can be moved in a natural fashion through space and time. However, the entire inbetween problem is not solved.

Inbetween processing also includes interpolation of the form of the objects in the intermediate frames. Some methods have been developed, including one by Burtnyk and Wein [BW76]. The authors develop a skeleton for a motion by choosing a polygon that describes the principal shape of a two-dimensional figure (or part of the figure) as well as the vicinity of the shape. The figure is represented in a coordinate system based on the skeleton. The intermediate processing proceeds by interpolating the characteristics of the skeleton between the key frames. A similar technique can also be transferred to the three-dimensional domain. In general, the process of interpolation between key frames is a complex problem.

6.1.4 Changing Colors

To process color changes, computer-based animation uses the Color Look-Up Table (CLUT) or Look-Up Table (LUT) of the graphics memory and the double buffering method, whereby two parts of a frame are stored in different areas of graphic memory. The graphic memory is divided into two fields, each having half as many bits per pixel as the whole graphic memory.

The animation is generated by manipulating the LUT. The simplest method is to cyclically change the colors of the LUT, thereby changing the colors of different parts of an image. Performing LUT animation is relatively fast. If a frame buffer of size 640×512 pixels uses eight bits per pixel for color, then an image contains 320 Kbit of data. Transferring a new image into the frame buffer, which takes place every 1/30 of a second, would require a bandwidth of over 9 Mbit/s. On the other hand, new LUT values can be sent very quickly since LUTs typically contain on the order of a few hundred to a thousand bytes.

6.2 Specification of Animations

Various languages exist to describe animations and new formal specification are currently being researched and further developed.

These specifications can be divided into the following three categories:

- Linear-List Notations

  In linear list notation each event in an animation is described by a beginning frame number, an end frame number, and an action (*event*) that is to be performed. Actions typically accept input parameters in the form of an instruction such as the following:

  `42, 53, B, ROTATE "PALM", 1, 30.`

  This instruction means that between frames 42 and 53, the object denoted `PALM` should be rotated 30 degrees around axis 1. A table determines the rotation in each individual frame, allowing for animations with either uniform or accelerated movement [FDFH92].

Many other linear list notations have been developed. Other notations are supersets of linear lists, an example being Scefo (SCEne Format) [Str88], which also includes the specification of groups and object hierarchies as well as transformation abstractions (so-called *actions*) by using constructs from high-level programming languages.

- High-Level Programming Language Notations

Another way to describe animations is by embedding animation control in a general-purpose programming language. The values of variables in the language can then be used as parameters for animation routines.

ASAS is an example of such a language [Rei82] based on an extension of LISP. The language's primitives include vectors, colors, polygons, surfaces, groups, points of view, subworlds, and aspects of lighting. ASAS also includes a large collection of geometric transformations that operate on objects. The following ASAS program fragment describes an animated sequence in which an object called my-cube is rotated while the camera pans. This fragment is evaluated for each frame in order to generate the entire sequence.

```
(grasp my-cube); Make the cube the current object
(cw 0.05); Small clock-wise rotation
(grasp camera); Make the camera the current object
(right panning-speed); Move it to the right
```

- Graphical Languages

A problem with traditional, textual programming languages is that graphical actions cannot be easily visualized by examining scripts. Graphical animation languages describe animations in a more visual fashion. Such languages are used to name and edit the changes taking place simultaneously in an animation and to visualize the effects created. The description of actions to be carried out is done using visual paradigms. GENESYS [Bae69], DIAL [FSB82] and the S-Dynamics System [Inc85] are examples of such systems.

6.3 Methods of Controlling Animation

Animation control is independent of the language used to describe animation. There are various techniques for controlling animation.

6.3.1 Explicitly Declared Control

Explicit control is the simplest type of animation control. In explicit control, the animator provides a description of all events that could occur in an animation. This can be done by specifying simple transformations—such as scalings, translations, and rotations—or by specifying key frames and methods for interpolating between them.

Interpolation can be specified either explicitly or, in an interactive system, through direct manipulation with a mouse, joystick, data glove, or other input device. An example of this type of control is the BBOP system [Ste83].

6.3.2 Procedural Control

Procedural control is based on communication among different objects whereby each object obtains knowledge about the static or dynamic properties of other objects. This information can be used to verify that objects move in a consistent fashion. In particular, in systems that represent physical processes, the position of an object can influence the movement of other objects (for example, ensuring that balls cannot move through walls). In actor-based systems, individual actors can pass their positions along to others in order to influence their behavior.

6.3.3 Constraint-Based Control

Although some objects in the real world move along straight lines, this is not always the case. Many objects' movements are determined by other objects with which they come in contact. It is thus much simpler to specify an animation sequence using constraints (usually determined by the environment) instead of explicit control. Sutherland's Sketchpad [Sut63] and Borning's ThingLab [Bor79] are examples of systems using constraints for control.

Currently much work is underway on support for hierarchies of conditions and on determination of motion. Many of these efforts allow for the specification of constraints that take into account the dynamics of real bodies and the structural properties of their materials.

6.3.4 Control by Analyzing Live Action

By examining the motions of objects in the real world, one can animate the same movement by creating corresponding sequences of objects. Traditional animation uses rotoscoping. A film is made in which people or animals act out the parts of the performers in the animation. Afterwards, animators process the film, enhancing the background and replacing the human actors with the animated equivalents they have created.

Another such technique is to attach indicators to key points on the body of a human actor. The coordinates of the corresponding key points in an animated model can be calculated by observing the position of these indicators. An example of this sort of interaction mechanism is the data glove, which measures the position and orientation of the wearer's hand, as well as the flexion and hyperextension of each finger point.

6.3.5 Kinematic and Dynamic Control

Kinematics refers to the position and velocity of points. A kinematic description of a scene, for example, might say, "The cube is at the origin at time $t=0$. Thereafter it moves with constant acceleration in the direction (1 meter, 1 meter, 5 meters)."

In contrast, dynamics takes into account the physical laws that govern kinematics (for example, the Newtonian laws for the movement of large bodies, or the Euler-Lagrange equations for fluids). A particle moves with an acceleration proportional to the forces acting on it; the proportionality constant is the mass of the particle. A dynamic description of a scene might be: "At time $t=0$, the cube is at position (0 meter, 100 meter, 0 meter). The cube has a mass of 100 grams. The force of gravity acts on the cube." The natural reaction in a dynamic simulation is that the cube would fall.

6.4 Display of Animation

To display animations with raster systems, the animated objects (which may consist of graphical primitives, such as lines or polygons) must be scan-converted and stored as a pixmap in the frame buffer. A rotating object can be shown by displaying successive views from slightly different locations.

The scan-conversion must be done at least 10 (preferably 15 to 20) times per second in order to give a reasonably smooth visual effect; hence a new image must be generated in at most 100 ms. The actual scan-conversion of an object should take only a small portion of this time. For example, if the scan-conversion took 75 ms, only 25 ms remain to erase and redraw the complete object on the screen; this is not enough time, and the result is a distracting ghost effect.

"Double buffering" is used to avoid this problem. As an example, consider the display of a rotation animation [FDFH92]. Assuming that the two halves of the pixmap are $image_0$ and $image_1$, the process is as follows:

```
Load LUT to display values as background color
Scan-convert object into image₀
Load LUT to display only image₀
Repeat
    Scan-convert object into image₁
    Load LUT to display only image₁
    Rotate object data structure description
    Scan-convert object into image₀
    Load LUT to display only image₀
    Rotate object data structure description
Until (termination condition).
```

If rotating and scan-converting the object takes more than 100 ms, the animation is quite slow, but the transition from one image to the next appears to be instantaneous. Loading the LUT typically takes less than 1 ms [FDFH92].

6.5 Transmission of Animation

Animated objects can be represented symbolically using graphical objects or scan-converted pixmap images. Hence, the transmission of an animation may be performed using one of two approaches:

- The symbolic representation (e.g., circle) of an animation's objects (e.g., ball) is transmitted together with the operations performed on the object (e.g., *roll the ball*). The receiver displays the animation as described earlier. Since the byte size of such a representation is much smaller than a pixmap representation, the transmission time is short. However, the display time is longer since the receiver must generate the corresponding pixmaps.

  In this approach, the bandwidth (e.g. bytes/second) required to transmit an animation depends on (1) the size of the symbolic representation structure used to encode the animated object, (2) the size of the structure used to encode the operation command, and (3) the number of animated objects and operation commands sent per second.

- The pixmap representations of the animated objects are transmitted and displayed by the receiver. The transmission time is longer compared to the previous approach because of the size of the pixmap representation. However, the display time is shorter.

  The necessary bandwidth is at least proportional to the size of a single pixmap image and to the image repetition rate. These values are significantly higher than in the case of a symbolic representation.

6.6 Virtual Reality Modeling Language (VRML)

The Virtual Reality Modeling Language (VRML) is a format for describing three-dimensional interactive worlds and objects that can be used together with the World Wide Web. For example, VRML can be used to generate three-dimensional representations of complex scenes such as illustrations, product definitions, or Virtual Reality presentations.

The idea of a platform-independent standard for 3-D WWW applications originated in May 1994 at the "First International Conference on the World-Wide Web." Immediately afterwards, the Internet magazine *Wired* set up a mailing list to collect proposals and suggestions on the topic. Five months later, in October 1994, VRML 1.0 was presented at the "Second International Conference on the World-Wide Web." Two weeks after SIGGRAPH 95 (in August 1995), the VRML Architecture Group (VAG) was established. As technical development of the language progressed, the VAG focused increasingly on the standardization process. In January 1996, the VAG called for submission of proposals for VRML 2.0. Proposals submitted by Apple, the German National Research Center for Information Technology (GMD), IBM Japan, Microsoft, and Silicon Graphics, Inc. (SGI), were voted on in March 1996 over the Internet. The "Moving Worlds" proposal submitted by SGI was accepted by a majority and considered to be the working basis for VRML 2.0. VRML 2.0 was publicly presented on August 6, 1996 at SIGGRAPH 96.

VRML was adopted as international standard ISO/IEC 14772, prepared by Joint Technical Committee ISO/IEC JTC 1, the Information Technology Sub-Committee 24 (computer graphics and image processing) in cooperation with the VRML Architecture Group (VAG) and the VRML mailing list (www-vrml@wired.com). ISO/IEC 14772 is a single standard under the general title *Information Technology—Computer Graphics and Image Processing—Virtual Reality Modeling Language* [Org97].

VRML is capable of representing static and animated objects as well as hyperlinks to other media such as sound, motion pictures, and still pictures. Interpreters (browsers) for VRML are widely available for many different platforms, as are authoring tools for generating VRML files.

VRML is a model that allows for the definition of new objects as well a registration process that makes it possible for application developers to define common extensions to the base standard. Furthermore, there are mappings between VRML elements and generally used features of 3-D Application Programmer Interfaces (API) [ANM96].

There are three ways of navigating through a virtual world:

- WALK: Movement over the ground at eye-level (the ground lies in the *x-z* plane).
- FLY: Movement at any height.
- EXAMINE: Rotating an object in order to examine it more closely during design.

The following color animation is an example of a VRML animation:

```
Color interpolator
This example interpolates in a 10-second long cycle
from red to green to blue
DEF myColor ColorInterpolator{
   key        [   0.0,     0.5,     1.0 ]
   keyValue   [ 1 0 0,   0 1 0,   0 0 1 ] # red, green, blue
}
DEF myClock TimeSensor{
   cycleInterval 10.0     # 10 second animation
   loop          TRUE     # animation in endless loop
}
ROUTE myClock.fraction_changed TO myColor.set_fraction
```

A more complex example demonstrates the capability of VRML to describe interactive animations of 3-D objects as well as camera parameters:

```
Elevator

Group {
  children [
    DEF ETransform Transform {
      children [
        DEF EViewpoint Viewpoint { }
        DEF EProximity ProximitySensor { size 2 2 2 }
        <Geometry of the elevator: a unit cube
        at the origin with a door>
      ]
    }
  ]
}
DEF ElevatorPI PositionInterpolator {
  keys [ 0, 1 ]
  values [ 0 0 0, 0 4 0 ] # a floor is 4 meters tall
}
DEF TS TimeSensor { cycleInterval 10 } # 10 second travel time
DEF S Script {
  field SFNode viewpoint USE EViewpoint
  eventIn SFBool active
  eventIn SFBool done
  eventOut SFTime start
  behavior "elevator.java"
}
ROUTE EProximity.enterTime TO TS.startTime
ROUTE TS.isActive TO EViewpoint.bind
ROUTE TS.fraction_changed TO ElevatorPI.set_fraction
ROUTE ElevatorPI.value_changed TO ETransform.set_translation
```

This example of a camera animation implements an elevator that facilitates access to a two-dimensional building that has many floors. It is assumed that the elevator is already on the ground floor and doesn't have to be moved there first. In order to go up, the user must enter the elevator. A so-called proximity sensor triggers upon entry, automatically starting the elevator. The control buttons are located outside of the elevator.

Further details about VRML can be found in [ANM96].

CHAPTER 7

Data Compression

In comparison to the text medium, video frames have high storage requirements. Audio and particularly video pose even greater demands in this regard. The data rates needed to process and send continuous media are also considerable. This chapter thus covers the efficient compression of audio and video, which will be explained first in theory and subsequently in terms of compression standards.

7.1 Storage Space

Uncompressed graphics, audio, and video data require substantial storage capacity, which is not possible in the case of uncompressed video data, even given today's CD and DVD technology. The same is true for multimedia communications. Data transfer of uncompressed video data over digital networks requires that very high bandwidth be provided for a single point-to-point communication. To be cost-effective and feasible, multimedia systems must use compressed video and audio streams.

Most compression methods address the same problems, one at a time or in combination. Most are already available as products. Others are currently under development or are only partially completed (see also [SPI94]). While fractal image compression [BH93] may be important in the future, the most important compression techniques in use today are JPEG [Org93, PM93, Wal91] for single pictures, H.263 ($p \times 64$) [Le 91, ISO93b] for video, MPEG [Lio91, ITUC90] for video and audio, as well as proprietary techniques such as QuickTime from Apple and Video for Windows from Microsoft.

In their daily work, developers and multimedia experts often need a good understanding of the most popular techniques. Most of today's literature, however, is either too comprehensive or is dedicated to just one of the above-mentioned compression techniques, which is then described from a very narrow point of view. In this chapter, we compare the most important techniques—JPEG, H.263, and MPEG—in order to show their advantages, disadvantages, their similarities and differences, as well as their suitability for today's multimedia systems (for further analysis, see [ES98]). First, the motivation for the use of compression techniques will be illustrated. Subsequently, requirements for compression techniques will be derived. Section 7.3 covers source, entropy, and hybrid coding. Sections 7.5 through 7.7 provide details about JPEG, H.263, and MPEG. Section 7.8 explains the basics of fractal compression.

7.2 Coding Requirements

Images have considerably higher storage requirements than text, and audio and video have still more demanding properties for data storage. Moreover, transmitting continuous media also requires substantial communication data rates. The figures cited below clarify the qualitative transition from simple text to full-motion video data and demonstrate the need for compression. In order to be able to compare the different data storage and bandwidth requirements of various visual media (text, graphics, images, and video), the following specifications are based on a small window of 640×480 pixels on a display. The following holds always:

1 kbit = 1000 bit

1 Kbit = 1024 bit

1 Mbit = 1024×1024 bit

1. For the representation of the text medium, two bytes are used for every 8×8 pixel character.

    Character per screen page = $\frac{640 \times 480}{8 \times 8}$ = 4,800

    Storage required per screen page = 4,800 × 2 byte = 9,600 byte = 9.4 Kbyte

2. For the representation of vector images, we assume that a typical image consists of 500 lines [BHS91]. Each line is defined by its coordinates in the x direction and the y direction, and by an 8-bit attribute field. Coordinates in the x direction require 10 bits ($\log_2(640)$), while coordinates in the y direction require 9 bits ($\log_2(480)$).

    Bits per line = 9 bits + 10 bits + 9 bits + 10 bits + 8 bits = 46 bits

    Storage required per screen page = $500 \times \frac{46}{8}$ = 2,875 byte = 2.8 Kbyte

Coding Requirements

3. Individual pixels of a bitmap can be coded using 256 different colors, requiring a single byte per pixel.

   Storage required per screen page = 640×480×1 byte = 307,200 byte = 300 Kbyte

The next examples specify continuous media and derive the storage required for one second of playback.

1. Uncompressed speech of telephone quality is sampled at 8 kHz and quantized using 8 bit per sample, yielding a data stream of 64 Kbit/s.

   Required storage space/s = $\frac{64 \text{ Kbit/s}}{8 \text{ bit/byte}} \times \frac{1 \text{ s}}{1{,}024 \text{ byte/Kbyte}}$ = 8 Kbyte

2. An uncompressed stereo audio signal of CD quality is sampled at 44.1 kHZ and quantized using 16 bits.

   Data rate = $2 \times \frac{44{,}100}{\text{s}} \times \frac{16 \text{ bit}}{8 \text{ bit/byte}}$ = 176,400 byte/s

   Required storage space/s = 176,400 byte/s × $\frac{1 \text{ s}}{1{,}024 \text{ byte/Kbyte}}$ = 172 Kbyte

3. A video sequence consists of 25 full frames per second. The luminance and chrominance of each pixel are coded using a total of 3 bytes.

   According to the European PAL standard, each frame consists of 625 lines and has a horizontal resolution of more than 833 pixels. The luminance and color difference signals are encoded separately and transmitted together using a multiplexing technique (4:2:2).

   According to CCIR 601 (studio standard for digital video), the luminance (Y) is sampled at 13.5 MHz, while chrominance (R–Y and B–Y) is sampled using 6.75 MHz. Samples are coded uniformly using 8 bits.

   Bandwidth = (13.5 MHz + 6.75 MHz + 6.75 MHz)×8 bit = 216×10^6 bit/s.

   Data rate = 640 × 480 × 25 × 3 byte/s = 23,040,000 byte/s

   Required storage sapce/s = $2{,}304 \times 10^4$ byte/s × $\frac{1 \text{ s}}{1{,}024 \text{ byte/Kbyte}}$ = 22,500 Kbyte

High-resolution television uses twice as many lines and an aspect ratio of 16:9, yielding a data rate 5.33 times that of current televisions.

These examples briefly illustrate the increased demands on a computer system in terms of required storage space and data throughput if still images and in particular continuous media are to be processed. Processing uncompressed video data streams in an integrated multimedia system requires storage space in the gigabyte range and buffer space in the megabyte range. The throughput in such a system can be as high as 140Mbit/s, which must also be transmitted over networks connecting systems (per unidirectional connection). This kind of data transfer rate is not realizable with today's technology, or in the near future with reasonably priced hardware.

However, these rates can be considerably reduced using suitable compression techniques [NH88, RJ91], and research, development, and standardization in this area have progressed rapidly in recent years [ACM89, GW93]. These techniques are thus an essential component of multimedia systems.

Several compression techniques for different media are often mentioned in the literature and in product descriptions:

- JPEG (Joint Photographic Experts Group) is intended for still images.
- H.263 (H.261 $p \times 64$) addresses low-resolution video sequences. This can be complemented with audio coding techniques developed for ISDN and mobile communications, which have also been standardized within CCITT.
- MPEG (Moving Picture Experts Group) is used for video and audio compression.

Compression techniques used in multimedia systems are subject to heavy demands. The quality of the compressed, and subsequently decompressed, data should be as good as possible. To make a cost-effective implementation possible, the complexity of the technique used should be minimal. The processing time required for the decompression algorithms, and sometimes also the compression algorithms, must not exceed certain time spans. Different techniques address requirements differently (see, for example, the requirements of [Org93]).

One can distinguish between requirements of "dialogue" mode applications (e.g. videoconferencing or picture transmission) and "retrieval" mode applications (e.g. audiovisual information systems, where a human user retrieves information from a multimedia database).

Compression techniques like $p \times 64$, with its symmetric processing expenditure for compression and decompression and its strict delay limits, are better suited to dialogue applications. Other techniques, such as MPEG-1, are optimized for use in retrieval applications at the expense of considerable effort during compression.

The following requirement applies to compression techniques used for dialogue mode applications:

- The end-to-end delay for a technique used in a dialogue system should not exceed 150ms for compression and decompression alone. To support an easy, natural dialogue, ideally compression or decompression by itself should introduce an additional delay of no more than 50ms. The overall end-to-end delay additionally comprises delay in the network, communications protocol processing in the end system, and data transfer to and from the respective input and output devices.

The following requirements apply to compression techniques used for *re*trieval mode applications:

- Fast forward and fast rewind with simultaneous display (or playback) of the data should be possible, so that individual passages can be sought and found quicker.
- Random access to single images or audio passages in a data stream should be possible in less than half a second. This access should be faster than a CD-DA system in order to maintain the interactive character of the retrieval application.
- Decompression of images, video, or audio passages should be possible without interpreting all preceding data. This allows random access and editing.

The following requirements apply to compression techniques used for either dialogue or retrieval mode applications:

- To support display of the same data in different systems, it is necessary to define a format that is independent of frame size and video frame rate.
- Audio and video compression should support different data rates (at different quality) so that the data rate can be adjusted to specific system conditions.
- It should be possible to precisely synchronize audio and video. It should also be possible for a system program to synchronize with other media.
- It should be possible to implement an economical solution today either in software or using at most a few VLSI chips.
- The technique should enable cooperation among different systems. It should be possible for data generated on a local multimedia system to be reproduced on other systems. This is relevant, for example, in the case of course materials distributed on CDs. This allows many participants to read the data on their own systems, which may be made by different manufacturers. Also, since many applications exchange multimedia data between systems over communications networks, the compression techniques must be compatible. This can be assured with *de jure* (e.g. ITU, ISO, or ECMA) and/or *de facto* standards.

Different compression schemes take these requirements into account to a varying extent.

Coding Type	Basis	Technique
Entropy Coding	Run-length Coding	
	Huffman Coding	
	Arithmetic Coding	
Source Coding	Prediction	DPCM
		DM
	Transformation	FFT
		DCT
	Layered Coding (according to importance)	Bit Position
		Subsampling
		Subband Coding
	Vector Quantization	
Hybrid Coding	JPEG	
	MPEG	
	H.263	
	Many proprietary systems	

Table 7-1 Overview of some coding and compression techniques.

7.3 Source, Entropy, and Hybrid Coding

Compression techniques can be categorized as shown in Table 7-1. We distinguish among three types of coding: entropy, source, and hybrid coding. Entropy coding is a lossless process, while source coding is often lossy. Most multimedia systems use hybrid techniques; most are only combinations of entropy and source coding, without any new processing algorithms.

7.3.1 Entropy Coding

Entropy coding can be used for different media regardless of the medium's specific characteristics. The data to be compressed are viewed as a sequence of digital data values, and their semantics are ignored. It is lossless because the data prior to encoding is identical to the data after decoding; no information is lost. Thus run-length encoding, for example, can be used for compression of any type of data in a file system, for example, text, still images for facsimile, or as part of motion picture or audio coding.

7.3.2 Source Coding

Source coding takes into account the semantics of the information to be encoded [SGC90]. The degree of compression attainable with this often lossy technique depends on the medium. In the case of lossy compression, a relation exists between the uncoded data and the decoded data; the data streams are similar but not identical. The characteristics of the medium can be exploited. In the case of speech, a considerable reduction of the amount of data can be achieved by transforming the time-dependent signal into the frequency domain, followed by an encoding of the formants (see Chapter 3 regarding audio).

Formants are defined as the maxima in the voice spectrum. Generally, five formants along with the voice fundamental frequency suffice for a very good reconstruction of the original signal, although formant tracking has been shown to be a problem with this form of speech analysis (see Chapter 3).

In the case of still images, spatial redundancies can be used for compression through a content prediction technique. Other techniques perform a transformation of the spatial domain into the two-dimensional frequency domain by using the cosine transform. Low frequencies define the average color, and the information of higher frequencies contains the sharp edges. Hence, low frequencies are much more important than the higher frequencies, a feature that can be used for compression.

Table 7-1 shows only a sampling of all coding and compression techniques. The emphasis is on the algorithms that are most important for multimedia systems and on their properties. To better understand the hybrid schemes, we consider a set of typical processing steps common to all techniques (entropy, source, and hybrid).

7.3.3 Major Steps of Data Compression

Figure 7-1 shows the typical sequence of operations performed in the compression of still images and video and audio data streams. The following example describes the compression of one image:

1. The preparation step (here picture preparation) generates an appropriate digital representation of the information in the medium being compressed. For example, a picture might be divided into blocks of 8×8 pixels with a fixed number of bits per pixel.
2. The processing step (here picture processing) is the first step that makes use of the various compression algorithms. For example, a transformation from the time domain to the frequency domain can be performed using the Discrete Cosine Transform (DCT). In the case of interframe coding, motion vectors can be determined here for each 8×8 pixel block.
3. Quantization takes place after the mathematically exact picture processing step. Values determined in the previous step cannot and should not be processed with

full exactness; instead they are quantized according to a specific resolution and characteristic curve. This can also be considered equivalent to the µ-law and *A*-law, which are used for audio data [JN84]. In the transformed domain, the results can be treated differently depending on their importance (e.g., quantized with different numbers of bits).

4. Entropy coding starts with a sequential data stream of individual bits and bytes. Different techniques can be used here to perform a final, lossless compression. For example, frequently occurring long sequences of zeroes can be compressed by specifying the number of occurrences followed by the zero itself.

Picture processing and quantization can be repeated iteratively, such as in the case of Adaptive Differential Pulse Code Modulation (ADPCM). There can either be "feedback" (as occurs during delta modulation), or multiple techniques can be applied to the data one after the other (like interframe and intraframe coding in the case of MPEG). After these four compression steps, the digital data are placed in a data stream having a defined format, which may also integrate the image starting point and type of compression. An error correction code can also be added at this point.

Figure 7-1 shows the compression process applied to a still image; the same principles can also be applied to video and audio data.

Figure 7-1   Major steps of data compression.

Decompression is the inverse process of compression. Specific coders and decoders can be implemented very differently. Symmetric coding is characterized by comparable costs for encoding and decoding, which is especially desirable for dialogue applications. In an asymmetric technique, the decoding process is considerably less costly than the coding process. This is intended for applications where compression is performed once and decompression takes place very frequently, or if the decompression must take place very quickly. For example, an audio-visual course module is produced once, but subsequently decoded by the many students who use it. The main requirement is real-time decompression. An asymmetric technique can be used to increase the quality of the compressed images.

The following section discusses some basic compression techniques. Subsequent sections describe hybrid techniques frequently used in the multimedia field.

7.4 Basic Compression Techniques

The hybrid compression techniques often used on audio and video data in multimedia systems are themselves composed of several different techniques. For example, each technique listed in Table 7-1 employs entropy coding (in the form of variations of run-length coding and/or a statistical compression technique).

The simplest techniques are based on interpolation, whereby it is possible to make use of properties of the human eye or ear. For example, the eye is more sensitive to changes in brightness than to color changes. Therefore, instead of dividing an image into RGB (red, green, blue) components, a YUV representation can be used (see Chapter 5). The U and V components can then be sampled with lower horizontal and vertical resolution, a technique called subsampling.

7.4.1 Run-Length Coding

Data often contains sequences of identical bytes. By replacing these repeated byte sequences with the number of occurrences, a substantial reduction of data can be achieved. This is known as run-length coding. A special marker M is needed in the data that does not occur as part of the data stream itself. This M-byte can also be realized if all 256 possible bytes can occur in the data stream by using byte stuffing. To illustrate this, we define the exclamation mark to be the M-byte. A single occurrence of an exclamation mark is interpreted as the M-byte during decompression. Two consecutive exclamation marks are interpreted as an exclamation mark occurring within the data.

The M-byte can thus be used to mark the beginning of a run-length coding. The technique can be described precisely as follows: if a byte occurs at least four times in a row, then the number of occurrences is counted. The compressed data contain the byte, followed by the M-byte and the number of occurrences of the byte. Remembering that we are compressing at least four consecutive bytes, the number of occurrences can be offset by -4. This allows the compression of between four and 258 bytes into only three bytes. (Depending on the algorithm, one or more bytes can be used to indicate the length; the same convention must be used during coding and decoding.)

In the following example, the character c occurs eight times in a row and is compressed to the three characters c!4:

Uncompressed data: ABCCCCCCCCDEFGGG

Run-length coded: ABC!4DEFGGG

7.4.2 Zero Suppression

Run-length coding is a generalization of zero suppression, which assumes that just one symbol appears particularly often in sequences. The blank (space) character in text is such a symbol; single blanks or pairs of blanks are ignored. Starting with sequences of three bytes, they are replaced by an M-byte and a byte specifying the number of blanks in the sequence. The number of occurrences can again be offset (by −3). Sequences of between three and 257 bytes can thus be reduced to two bytes. Further variations are tabulators used to replace a specific number of null bytes and the definition of different M-bytes to specify different numbers of null bytes. For example, an M5-byte could replace 16 null bytes, while an M4-byte could replace 8 null bytes. An M5-byte followed by an M4-byte would then represent 24 null bytes.

7.4.3 Vector Quantization

In the case of vector quantization, a data stream is divided into blocks of *n* bytes each ($n>1$). A predefined table contains a set of patterns. For each block, the table is consulted to find the most similar pattern (according to a fixed criterion). Each pattern in the table is associated with an index. Thus, each block can be assigned an index. Such a table can also be multidimensional, in which case the index will be a vector. The corresponding decoder has the same table and uses the vector to generate an approximation of the original data stream. For further details see [Gra84] for example.

7.4.4 Pattern Substitution

A technique that can be used for text compression substitutes single bytes for patterns that occur frequently. This pattern substitution can be used to code, for example, the terminal symbols of high-level languages (begin, end, if). By using an M-byte, a larger number of words can be encoded–the M-byte indicates that the next byte is an index representing one of 256 words. The same technique can be applied to still images, video, and audio. In these media, it is not easy to identify small sets of frequently occurring patterns. It is thus better to perform an approximation that looks for the most similar (instead of the same) pattern. This is the above described vector quantization.

7.4.5 Diatomic Encoding

Diatomic encoding is a variation based on combinations of two data bytes. This technique determines the most frequently occurring pairs of bytes. Studies have shown that the eight most frequently occurring pairs in the English language are "E," "T," "TH," "A," "S," "RE," "IN," and "HE." Replacing these pairs by special single bytes that otherwise do not occur in the text leads to a data reduction of more than ten percent.

7.4.6 Statistical Coding

There is no fundamental reason that different characters need to be coded with a fixed number of bits. Morse code is based on this: frequently occurring characters are coded with short strings, while seldom-occurring characters are coded with longer strings. Such statistical coding depends how frequently individual characters or sequences of data bytes occur. There are different techniques based on such statistical criteria, the most prominent of which are Huffman coding and arithmetic coding.

7.4.7 Huffman Coding

Given the characters that must be encoded, together with their probabilities of occurrence, the Huffman coding algorithm determines the optimal coding using the minimum number of bits [Huf52]. Hence, the length (number of bits) of the coded characters will differ. The most frequently occurring characters are assigned to the shortest code words. A Huffman code can be determined by successively constructing a binary tree, whereby the leaves represent the characters that are to be encoded. Every node contains the relative probability of occurrence of the characters belonging to the subtree beneath the node. The edges are labeled with the bits 0 and 1.

The following brief example illustrates this process:

1. The letters *A, B, C, D,* and *E* are to be encoded and have relative probabilities of occurrence as follows:

   $p(A)=0.16$, $p(B)=0.51$, $p(C)=0.09$, $p(D)=0.13$, $p(E)=0.11$

2. The two characters with the lowest probabilities, *C* and *E,* are combined in the first binary tree, which has the characters as leaves. The combined probability of their root node *CE* is 0.20. The edge from node *CE* to *C* is assigned a 1 and the edge from *CE* to *C* is assigned a 0. This assignment is arbitrary; thus, different Huffman codes can result from the same data.

3. Nodes with the following relative probabilities remain:

   $p(A)=0.16$, $p(B)=0.51$, $p(CE)=0.20$, $p(D)=0.13$

   The two nodes with the lowest probabilities are *D* and *A*. These nodes are combined to form the leaves of a new binary tree. The combined probability of the root node *AD* is 0.29. The edge from *AD* to *A* is assigned a 1 and the edge from *AD* to *D* is assigned a 0.

   If root nodes of different trees have the same probability, then trees having the shortest maximal path between their root and their nodes should be combined first. This keeps the length of the code words roughly constant.

4. Nodes with the following relative probabilities remain:

   $p(AD)=0.29$, $p(B)=0.51$, $p(CE)=0.20$

   The two nodes with the lowest probabilities are *AD* and *CE*. These are combined into a binary tree. The combined probability of their root node *ADCE* is 0.49. The edge from *ADCE* to *AD* is assigned a 0 and the edge from *ADCE* to *CE* is assigned a 1.

5. Two nodes remain with the following relative probabilities:

   $p(ADCE)=0.49$, $p(B)=0.51$

These are combined to a final binary tree with the root node *ADCEB*. The edge from *ADCEB* to *B* is assigned a 1, and the edge from *ADCEB* to *ADCE* is assigned a 0.

6. Figure 7-2 shows the resulting Huffman code as a binary tree. The result is the following code words, which are stored in a table:

$w(A)=001$, $w(B)=1$, $w(C)=011$, $w(D)=000$, $w(E)=010$

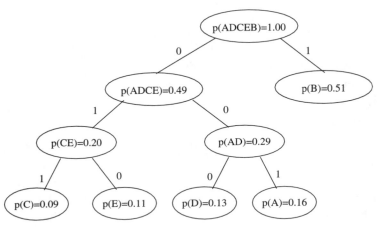

Figure 7-2 Example of a Huffman code represented as a binary tree.

Such a table could be generated for a single image or for multiple images together. In the case of motion pictures, a Huffman table can be generated for each sequence or for a set of sequences. The same table must be available for both encoding and decoding. If the information of an image can be transformed into a bit stream, then a Huffman table can be used to compress the data without any loss. The simplest way to generate such a bit stream is to code the pixels individually and read them line by line. Note that usually more sophisticated methods are applied, as described in the remainder of this chapter.

If one considers run-length coding and all the other methods described so far, which produce the same consecutive symbols (bytes) quite often, it is certainly a major objective to transform images and videos into a bit stream. However, these techniques also have disadvantages in that they do not perform efficiently, as will be explained in the next step.

7.4.8 Arithmetic Coding

Like Huffman coding, arithmetic coding is optimal from an information theoretical point of view (see [Lan84, PMJA88]). Therefore, the length of the encoded data is also minimal. Unlike Huffman coding, arithmetic coding does not code each symbol separately. Each symbol is instead coded by considering all prior data. Thus a data stream encoded in this fashion must always be read from the beginning. Consequently, random access is not possible. In practice, the average compression rates achieved by arithmetic and Huffman coding are similar [Sto88].

Differences in compression efficiency arise in special cases, for example in digitized graphics consisting primarily of a single (background) color. Arithmetic coding is better suited than Huffman coding in this case. This is always the case if symbols occur in the input data stream with such a frequency that they have a very small information content. These can be encoded using less than one bit, whereas in Huffman coding each symbol requires at least one bit.

7.4.9 Transformation Coding

Transformation coding pursues a different approach. Data is transformed into another mathematical domain that is more suitable for compression. An inverse transformation must exist. The simplest example is the Fourier transform, which transforms data from the time domain into the frequency domain. Other examples include the Walsh, Hadamard, Haar, and Slant transforms. However, the transformed data have no major advantages with respect to a subsequent compression. The most effective transformations for data reduction are the Discrete Cosine Transform (DCT), described in Section 7.5.2, and the Fast Fourier Transform (FFT).

7.4.10 Subband Coding

Unlike transformation coding, which transforms all data into another domain, selective frequency transformation (subband coding) considers the signal only in predefined regions of the spectrum, such as frequency bands. The number of bands is a crucial quality criterion. This technique is well suited for the compression of speech.

7.4.11 Prediction or Relative Coding

Instead of compressing single bytes or sequences of bytes, differential encoding can be used. This is also known as prediction or relative coding. For example, if characters in a sequence are clearly different from zero, but do not differ much amongst themselves, then calculating differences from the respective previous value could be profitable for compression. The following examples explain this technique for different media:

- For still images, edges yield large difference values, while areas with similar luminance and chrominance yield small values. A homogeneous area is characterized by a large number of zeroes, which could be further compressed using run-length coding.

- For video, performing relative coding in the time domain leads to an encoding of differences from the previous image only. In a newscast or a video telephony application, there will be a large number of zero bytes, because the background of successive images does not change very often. Motion compensation can also be performed here (see e.g. [PA91]). Blocks of, for example, 16×16 pixels are compared with each other in successive pictures. In the case of a car moving from left to right, an area in the current image would be most similar to an area lying further to the left in the previous image. This motion can be coded as a vector.
- Audio techniques often apply Differential Pulse Code Modulation (DPCM) to a sequence of PCM-coded samples (see e.g. [JN84]). This technique requires a linear characteristic curve for quantization. It is thus not necessary to store the whole number of bits for each sample. It is sufficient to represent the first PCM-coded sample as a whole and all following samples as the difference from the previous one.

7.4.12 Delta Modulation

Delta Modulation is a modification of DPCM where difference values are encoded with exactly one bit, which indicates whether the signal increases or decreases. This leads to an inaccurate coding of steep edges. This technique is particularly profitable if the coding does not depend on 8-bit grid units. If the differences are small, then a much smaller number of bits is sufficient. Difference encoding is an important feature of all techniques used in multimedia systems. Section 7.5.2 describes other "delta" methods that can be applied to images.

7.4.13 Adaptive Compression Techniques

Most of the compression techniques described so far take advantage of already known characteristics of the data to be compressed (e.g., frequently occurring sequences of bytes or the probability of occurrence of individual bytes). An atypical sequence of characters results in a poor compression. However, there are also adaptive compression techniques, which can adjust themselves to the data being compressed. This adaptation can be implemented in different ways:

- We illustrate the first technique with the following example, which assumes a coding table has been generated in advance (e.g., as per Huffman). For each symbol to be encoded, the table contains the corresponding code word and, in an additional column, a counter. All entries' counters are initialized to zero at the beginning. For the first symbol to be encoded, the coder determines the code word according to the table. Additionally, the coder increments the counter by one. The symbols and their respective counters are then sorted in decreasing order by counter value. The order of the code words is not changed. The most frequently occurring symbols are now at the beginning of the table, since they have the highest counter values. Symbols with the highest frequency count are thus always encoded with the shortest code words.

- Another adaptive compression technique is Adaptive DPCM (ADPCM), a generalization of DPCM. For simplicity, this is often called DPCM.

Here, difference values are encoded using only a small number of bits. With DPCM, either coarse transitions would be coded correctly (the DPCM-encoded bits are used represent bits with a higher significance), or fine transitions are coded exactly (if the bits are used to represent bits with less significance). In the first case, the resolution of low audio signals would not be sufficient, and in the second case a loss of high frequencies would occur.

ADPCM adapts to the significance of the data stream. The coder divides the DPCM sample values by an appropriate coefficient and the decoder multiplies the compressed data by the same coefficient, thus changing the step size of the signal. The coder of the DPCM-encoded signal adjusts the value of the coefficient.

A signal with a large portion of high frequencies will result in frequent, very high DPCM values. The coder will select a high value for the coefficient. The result is a very rough quantization of the DPCM signal in passages with steep edges. Low-frequency portions of such passages are hardly considered at all.

In the case of a signal with steady, relatively low DPCM values, that is, with a small portion of high frequencies, the coder will choose a small coefficient. This ensures good resolution for the dominant, low frequency portion of the signal. If high frequencies suddenly occur in such a passage, a signal distortion in the form of a slope overload occurs. The greatest possible change that can be represented by an ADPCM value using the available number of bits and the current step size is not enough to represent the new DPCM value. The jump in the PCM signal will be faded.

Changes in the adaptively set coefficient can be explicitly inserted in the compressed data by the encoder. Alternatively, the decoder can calculate the coefficients itself from an ADPCM-coded data stream. This predictor is rated so as to minimize errors in the data stream. Note that the definition of an error and the associated predictor rating depends on the medium and is ideally trivial.

An audio signal with frequently changing portions of extreme low or high frequencies is generally not suited for ADPCM coding. For telephone applications, the ITU has standardized a version of the ADPCM technique using 32 Kbit/s that is based on four bits per difference value and a sampling frequency of 8 kHz.

7.4.14 Other Basic Techniques

Apart from the compression techniques described earlier, some additional well known techniques are used today:

- Video compression techniques often use Color Look-Up Tables (see Section 5.1.2). This technique is used in distributed multimedia systems, for example in [LE91, LEM92].
- A simple technique for audio is silence suppression, whereby data is only encoded if the volume exceeds a certain threshold.

ITU incorporates some of the basic audio coding techniques in the G.700 series of standards: G.721 defines PCM coding for 3.4 kHz quality over 64 Kbit/s channels, and G.728 defines 3.4 kHz quality over 16 Kbit/s channels. See [ACG93] for a detailed description of various audio coding techniques.

The following sections describe the most important work in the standardization bodies concerning image and video coding. In the framework of ISO/IECJTC1/SC2/WG8, four subgroups were established in May 1988: JPEG (Joint Photographic Experts Group), working on coding of still images; JBIG (Joint Bi-Level Image Experts Group), working on the progressive processing of bi-level coding algorithms; CGEG (Computer Graphics Experts Group), working on coding principles; and MPEG (Moving Picture Experts Group), working on the coded representation of motion video. The next section presents the results of the JPEG activities.

7.5 JPEG

Since June 1982, Working Group 8 (WG8) of ISO has been working on standards for the compression and decompression of still images [HYS88]. In June 1987, ten different techniques for coding color and gray-scaled still images were presented. These ten were compared, and three were analyzed further. An adaptive transformation coding technique based on the Discrete Cosine Transform (DCT) achieved the best subjective results [LMY88, WVP88]. This technique was then further developed with consideration paid to the other two methods. The coding known as JPEG (Joint Photographic Experts Group) is a joint project of ISO/IECJTC1/SC2/WG10 and Commission Q.16 of CCITT SGVIII. Hence the "J" (from "Joint") in JPEG—ISO together with CCITT. In 1992, JPEG became an ISO International Standard (IS) [Org93].

JPEG applies to color and gray-scaled still images [LOW91, MP91, Wal91]. Video sequences can also be handled through a fast coding and decoding of still images, a technique known as Motion JPEG. Today, implementations of parts of JPEG are already available, either as software-only packages or using special-purpose hardware support. It should be noted that as yet, most products support only the absolutely necessary algorithms. The only part of JPEG currently commercially available is the base mode with certain processing restrictions (limited number of image components and color coding).

7.5.0.1 Requirements

In order to ensure the widespread distribution and application of JPEG, the following requirements were established and fulfilled [Wal91]:

- The standard should be independent of image size.
- It should be applicable to any image aspect ratio and any pixel aspect ratio.
- The color space and the number of colors used should be independent of one another.
- Image content may be of any complexity, with any statistical characteristics.
- The JPEG standard should be start-of-the-art (or near) regarding the compression factor and image quality.
- The processing complexity should permit a software solution to run on a large number of available general-purpose processors, and should be drastically reduced with the use of special-purpose hardware.
- Sequential (line by line) and progressive (successive refinement of the whole image) decoding should both be possible. A lossless, hierarchical coding of the same image with different resolutions should also be possible.

The user can select parameters to trade off the quality of the reproduced image, the compression processing time, and the size of the compressed image.

7.5.0.2 JPEG Overview

Applications do not have to include both an encoder and a decoder. In many applications only one of them is needed. The encoded data stream has a fixed interchange format that includes encoded image data, as well as the chosen parameters and tables of the coding process, enabling decoding. If there is a common context between the coder and the decoder (e.g., if the coder and decoder are parts of the same application), then there can be an abbreviated interchange format. This format includes few if any of the required tables (Appendix A in [Org93] describes this format in detail). The interchange format includes all information required during decoding, if this information is not available as part of the common context. In the regular, non-abbreviated mode, the interchange format includes all information required to decode without prior knowledge of the coding process.

Figure 7-3 outlines the fundamental steps of JPEG compression in accordance with the general scheme illustrated in Figure 7-1. JPEG defines several image compression modes by selecting different combinations of these steps.

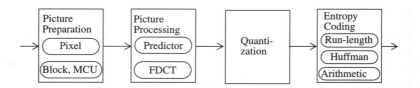

Figure 7-3 Steps of the JPEG compression technique: summary of the different modes.

7.5.0.3 JPEG Modes

JPEG defines four modes, which themselves include additional variations:

- The lossy, sequential DCT-based mode (baseline process, base mode) must be supported by every JPEG decoder.
- The expanded lossy, DCT-based mode provides a set of further enhancements to the base mode.
- The lossless mode has a low compression ratio and allows a perfect reconstruction of the original image.
- The hierarchical mode accommodates images of different resolutions by using algorithms defined for the other three modes.

The baseline process uses the following techniques: block, Minimum Coded Unit (MCU), FDCT, run-length, and Huffman, which are explained in this section together with the other modes. Image preparation is first presented for all modes. The picture processing, quantization, and entropy encoding steps used in each mode are then described separately for each mode.

7.5.1 Image Preparation

For image preparation, JPEG specifies a very general image model that can describe most commonly used still image representations. For instance, the model is not based on three image components with 9-bit YUV coding and a fixed numbers of lines and columns. The mapping between coded color values and the colors they represent is also not coded. This fulfills the JPEG requirement of independence from image parameters such as image size or the image and pixel aspect ratios.

An image consists of at least one and at most $N=255$ components or planes, as shown on the left side of Figure 7-4. These planes can be assigned to individual RGB (red, green, blue) colors, or to the YIQ or YUV signals, for example.

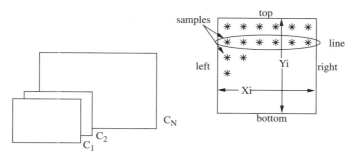

Figure 7-4 Uncompressed digital image as per [Org93].

Each component is a rectangular array $X_i \times Y_i$ of pixels (the samples). Figure 7-5 shows an image with three planes, each with the same resolution.

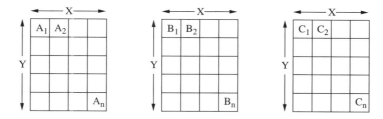

Figure 7-5 Example of JPEG image preparation with three components having the same resolution.

The resolution of the individual components may be different. Figure 7-6 shows an image with three planes where the second and third planes have half as many columns as the first plane. A gray-scale image will, in most cases, consist of a single component, while an RGB color image will have three components with the same resolution (same number of lines $Y_1 = Y_2 = Y_3$ and the same number of columns $X_1 = X_2 = X_3$). In JPEG image preparation, YUV color images with subsampling of the chrominance components use three planes with $Y_1 = 4Y_2 = 4Y_3$ and $X_1 = 4X_2 = 4X_3$.

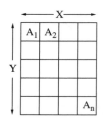 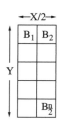 

Figure 7-6 Example of JPEG image preparation with three components having different resolutions.

Each pixel is represented by p bits with values in the range from 0 to 2^p-1. All pixels of all components of an image must have the same number of bits. The lossy modes of JPEG use a precision of either 8 or 12 bits per pixel. The lossless modes can use between 2 and 12 bits per pixel. If a JPEG application uses any other number of bits, the application itself must suitably transform the image to conform to the number of bits defined by the JPEG standard.

Instead of the values X_i and Y_i, the compressed data includes the values X (maximum of all X_i) and Y (maximum of all Y_i), as well as factors H_i and V_i for each plane. H_i and V_i represent the relative horizontal and vertical resolutions with respect to the minimal horizontal and vertical resolutions.

Let us consider the following example from [Org93]. An image has a maximum resolution of 512 pixels in both the horizontal and vertical directions and consists of three planes. The following factors are given:

```
Plane 0: H0 = 4, V0 = 1
Plane 1: H1 = 2, V1 = 2
Plane 2: H2 = 1, V2 = 1
```

This leads to:

```
X = 512, Y = 512, Hmax = 4 and Vmax = 2
Plane 0: X0 = 512, Y0 = 256
Plane 1: X1 = 256, Y1 = 512
Plane 2: X2 = 128, Y2 = 256
```

H_i and V_i must be integers between 1 and 4. This awkward-looking definition is needed for the interleaving of components.

In the image preparation stage of compression, the image is divided into data units. The lossless mode uses one pixel as one data unit. The lossy mode uses blocks of 8×8 pixels. This is a consequence of DCT, which always transforms connected blocks.

Up to now, the data units are usually prepared component by component and passed on in order to the following image processing. Within each component, the data units are processed from left to right, as shown in Figure 7-7. This is known as a noninterleaved data ordering.

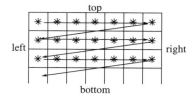

Figure 7-7 Noninterleaved processing order of data units when processing a single component as per [Org93].

Due to the finite processing speed of the decoder, processing of data units of different components may be interleaved. If the noninterleaved mode were used for a very high-resolution, RGB-encoded image, during rendering the display would first show only red, then red-green, and finally the correct colors.

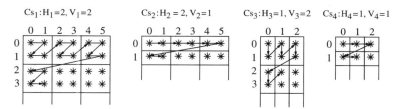

Figure 7-8 Interleaved processing order of data units.

Figure 7-8 shows an example with four components from [Org93]. Above each component, the corresponding values for H and V are shown. The first component has the highest resolution in both dimensions and the fourth component has the lowest resolution. The arrows within each component indicate the sampling order of individual data units.

Minimum Coded Units (MCUs) are built in the following order:

$$MCU_1 = d_{00}^1 d_{01}^1 d_{10}^1 d_{11}^1 d_{00}^2 d_{01}^2 d_{00}^3 d_{10}^3 d_{00}^4$$

$$MCU_2 = d_{02}^1 d_{03}^1 d_{12}^1 d_{13}^1 d_{02}^2 d_{03}^2 d_{01}^3 d_{11}^3 d_{01}^4$$

$$MCU_3 = d_{04}^1 d_{05}^1 d_{14}^1 d_{15}^1 d_{04}^2 d_{05}^2 d_{02}^3 d_{12}^3 d_{02}^4$$

$$MCU_4 = d_{20}^1 d_{21}^1 d_{30}^1 d_{31}^1 d_{10}^2 d_{11}^2 d_{20}^3 d_{30}^3 d_{10}^4$$

The data units of the first component are $C_{s1}: d_{00}^1 \ldots d_{31}^1$

The data units of the second component are $C_{s2}: d_{00}^2 \ldots d_{11}^2$

The data units of the third component are $C_{s3}: d_{00}^3 \ldots d_{30}^3$

The data units of the fourth component are $C_{s4}: d_{00}^4 \ldots d_{10}^4$

Interleaved data units of different components are combined into Minimum Coded Units. If all components have the same resolution ($X_i \times Y_i$), an MCU consists of exactly one data unit from each component. The decoder displays the image MCU by MCU. This allows for correct color presentation, even for partly decoded images.

In the case of different component resolutions, the construction of MCUs becomes more complex (see Figure 7-8). For each component, regions of data units, potentially with different numbers of data units, are constructed. Each component consists of the same number of regions. For example, in Figure 7-8 each component has six regions. MCUs are comprised of exactly one region from each component. The data units within a region are ordered from left to right and from top to bottom.

According to the JPEG standard, a maximum of four components can be encoded using the interleaved mode. This is not a limitation, since color images are generally represented using three components. Each MCU can contain at most ten data units. Within an image, some components can be encoded in the interleaved mode and others in the noninterleaved mode.

7.5.2 Lossy Sequential DCT-Based Mode

After image preparation, the uncompressed image samples are grouped into data units of 8×8 pixels, as shown in Figure 7-9; the order of these data units is defined by the MCUs. In this baseline mode, each sample is encoded using $p=8$ bit. Each pixel is an integer in the range 0 to 255.

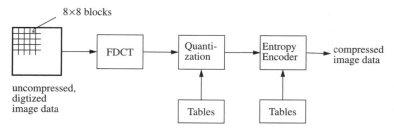

Figure 7-9 Steps of the lossy sequential DCT-based coding mode, starting with an uncompressed image after image preparation.

Figure 7-10 shows the compression and decompression process outlined here.

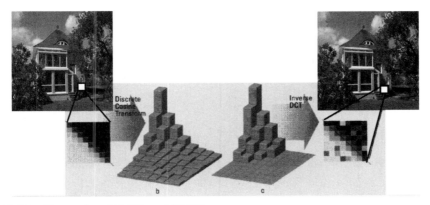

Figure 7-10 Example of DCT and IDCT.

7.5.2.1 Image Processing

The first step of image processing in the baseline mode (baseline process in [Org93]), as shown in Figure 7-9, is a transformation coding performed using the Discrete Cosine Transform (DCT) [ANR74, NP78]. The pixel values are shifted into the zero-centered interval $(-128, 127)$. Data units of 8×8 shifted pixel values are defined by S_{yx}, where x and y are in the range of zero to seven.

The following FDCT (Forward DCT) is then applied to each transformed pixel value:

$$S_{vu} = \frac{1}{4} c_u c_v \sum_{x=0}^{7} \sum_{y=0}^{7} S_{yx} \cos\frac{(2x+1)u\pi}{16} \cos\frac{(2y+1)v\pi}{16}$$

where: $c_u, c_v = \frac{1}{\sqrt{2}}$ for $u, v = 0$; otherwise $c_u, c_v = 1$

Altogether, this transformation must be carried out 64 times per data unit. The result is 64 coefficients S_{vu}. Due to the dependence of DCT on the Discrete Fourier Transform (DFT), which maps values from the time domain to the frequency domain, each coefficient can be regarded as a two-dimensional frequency.

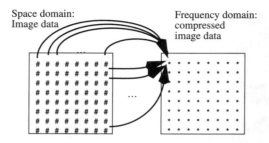

Figure 7-11 Relationship between the two-dimensional space and frequency domains.

The coefficient S_{00} corresponds to the portion where the frequency is zero in both dimensions. It is also known as the DC-coefficient (DC voltage portion) and determines the fundamental color of all 64 pixels of the data unit. The other coefficients are called AC-coefficients (analogous to the AC voltage portion). For instance, S_{70} represents the highest frequency that occurs in the horizontal direction, that is, the closest possible separation of vertical lines in the 8×8 data unit. S_{07} represents the highest frequency in the vertical dimension, that is, the closest possible separation of horizontal lines. S_{77} indicates the highest frequency appearing equally in both dimensions. Its absolute value is greatest if the original data unit contains as many squares as possible, that is, if it consists solely of 1×1 squares. Accordingly, for example, S_{33} will be maximal if a block consists of 16 squares of 4×4 pixels. Taking a closer look at the above formula, we recognize that the cosine expressions depend only upon x and u, or upon y and v, but not on S_{yx}. Therefore, these expressions represent constants that do not need to be recalculated over and over again. There are many effective DCT techniques and implementations [DG90, Fei90, Hou88, Lee84, LF91, SH86, VN84, Vet85].

For later reconstruction of the image, the decoder uses the Inverse DCT (IDCT). The coefficients S_{vu} must be used for the calculation:

$$s_{xy} = \frac{1}{4} \sum_{u=0}^{7} \sum_{v=0}^{7} c_u c_v S_{vu} \cos\frac{(2x+1)u\pi}{16} \cos\frac{(2y+1)v\pi}{16}$$

where $c_u, c_v = \frac{1}{\sqrt{2}}$ for $u, v = 0$; otherwise $c_u, c_v = 1$

If the FDCT, as well as the IDCT, could be calculated with full precision, it would be possible to reproduce the original 64 pixels exactly. From a theoretical point of view, DCT would be lossless in this case. In practice, precision is limited and DCT is thus lossy. The JPEG standard does not define any specific precision. It is thus possible that two different JPEG decoder implementations could generate different images as output of the same compressed data. JPEG merely defines the maximum allowable deviation.

Many images contain only a small portion of sharp edges; they consist mostly of areas of a single color. After applying DCT, such areas are represented by a small portion of high frequencies. Sharp edges, on the other hand, are represented as high frequencies. Images of average complexity thus consist of many AC-coefficients with a value of zero or almost zero. This means that subsequent entropy encoding can be used to achieve considerable data reduction.

7.5.2.2 Quantization

Image processing is followed by the quantization of all DCT coefficients; this is a lossy process. For this step, the JPEG application provides a table with 64 entries, one for each of the 64 DCT coefficients. This allows each of the 64 coefficients to be adjusted separately. The application can thus influence the relative significance of the different coefficients. Specific frequencies can be given more importance than others depending on the characteristics of the image material to be compressed. The possible compression is influenced at the expense of achievable image quality.

The table entries Q_{vu} are integer values coded with 8 bits. Quantization is performed according to the formula:

$$sq_{vu} = \text{round}\frac{S_{vu}}{Q_{vu}}$$

The greater the table entries, the coarser the quantization. Dequantization is performed prior to the IDCT according to the formula:

$$R_{vu} = Sq_{vu} \times Q_{vu}$$

Quantization and dequantization must use the same tables.

Figure 7-12 shows a greatly enlarged detail from Figure 7-16. The blocking and the effects of quantization are clearly visible. In Figure 7-12(b) a coarser quantization was performed to highlight the edges of the 8×8 blocks.

Figure 7-12 Quantization effect.

7.5.2.3 Entropy Encoding

During the next step, either the initial step of entropy encoding or preparation for the coding process, the quantized DC-coefficients are treated differently than the quantized AC-coefficients. The processing order of all coefficients is specified by the zig-zag sequence.

- The DC-coefficients determine the fundamental color of the data units. Since this changes little between neighboring data units, the differences between successive DC-coefficients are very small values. Thus each DC-coefficient is encoded by subtracting the DC-coefficient of the previous data unit, as shown in Figure 7-13, and subsequently using only the difference.

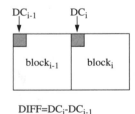

$DIFF=DC_i-DC_{i-1}$

Figure 7-13 Preparation of DCT DC-coefficients for entropy encoding. Calculation of the difference between neighboring values.

- The DCT processing order of the AC-coefficients using the zig-zag sequence as shown in Figure 7-14 illustrates that coefficients with lower frequencies (typically with higher values) are encoded first, followed by the higher frequencies (typically zero or almost zero). The result is an extended sequence of similar data bytes, permitting efficient entropy encoding.

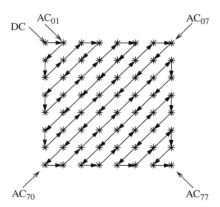

Figure 7-14 Preparation of DCT AC-coefficients for entropy encoding, in order of increasing frequency.

JPEG uses Huffman coding and arithmetic coding as entropy encoding methods. For the lossy sequential DCT-based base mode, discussed in this section, only Huffman encoding is allowed. In both methods, a run-length encoding of zero values is first applied to the quantized AC-coefficients. Additionally, non-zero AC-coefficients as well as the DC-coefficients are transformed into a spectral representation to further compress the data. The number of bits required depends on the value of each coefficient. Non-zero AC-coefficients are represented using between 1 and 10 bits. For the representation of DC-coefficients, a higher resolution of 1 bit to a maximum of 11 bits is used.

The result is a representation according to the ISO Intermediate Symbol Sequence Format, which basically alternates the following three pieces of information:

1. the number of subsequent coefficients with the value zero,
2. the number of bits used for the representation of the coefficient that follows, and
3. the value of the coefficient, represented using the specified number of bits.

An advantage of Huffman coding is that it can be used cost-free, since it is not protected by patents. A disadvantage is that the application must provide coding tables, since JPEG does not specify any. In the base mode, two different Huffman tables can be used, one for AC-coefficients and for DC-coefficients.

In sequential coding, the whole image is coded and decoded in a single run. Figure 7-15 shows an example of decoding with immediate presentation on the display. The picture is completed from top to bottom.

Figure 7-15 Example of sequential picture preparation, here using the lossy DCT-based mode.

7.5.3 Expanded Lossy DCT-Based Mode

Image preparation in this mode differs from the previously described mode in terms of the number of bits per pixel. This mode supports 12 bits per sample value in addition to 8 bits. The image processing is DCT-based and is performed analogously to the baseline DCT mode. For the expanded lossy DCT-based mode, JPEG defines progressive coding in addition to sequential coding. In the first decoding run, a very rough, unsharp image appears. This is refined during successive runs. An example of a very unsharp image is shown in Figure 7-16(a). It is substantially sharper in Figure 7-16(b), and in its correct resolution in Figure 7-16(c).

Figure 7-16 Progressive picture presentation: (a) first phase, very unsharp image; (b) second phase, unsharp image; (c) third phase, sharp image.

Progressive image presentation is achieved by expanding quantization. This is equivalent to layered coding. For this expansion, a buffer is added at the output of the quantizer that temporarily stores all coefficients of the quantized DCT. Progressiveness is achieved in two different ways:

- Using spectral selection, in the first run only the quantized DCT-coefficients of each data unit's low frequencies are passed on to the entropy encoding. Successive runs gradually process the coefficients of higher frequencies.
- In successive approximation, all of the quantized coefficients are transferred in each run, but individual bits are differentiated according to their significance. The most significant bits are encoded before the least significant bits.

Besides Huffman coding, arithmetic entropy coding can be used in the expanded mode. Arithmetic coding automatically adapts itself to the statistical characteristics of an image and thus requires no tables from the application. According to several publications, arithmetic encoding achieves around five to ten percent better compression than Huffman encoding. Other authors presuppose a comparable compression rate. Arithmetic coding is slightly more complex and its protection by patents must be considered (Appendix L in [Org93]).

In the expanded mode, four coding tables are available for the transformation of DC- and AC-coefficients. The simpler mode allows a choice of only two Huffman tables each for the DC- and AC-coefficients of an image. The expanded mode thus offers 12 alternative types of processing, as listed in Table 7-2.

Image Display	Bits per Sample Value	Entropy Coding
Sequential	8	Huffman coding
Sequential	8	Arithmetic coding
Sequential	12	Huffman coding
Sequential	12	Arithmetic coding
Progressive successive	8	Huffman coding
Progressive spectral	8	Huffman coding
Progressive successive	8	Arithmetic coding
Progressive spectral	8	Arithmetic coding
Progressive successive	12	Huffman coding
Progressive spectral	12	Huffman coding
Progressive successive	12	Arithmetic coding
Progressive spectral	12	Arithmetic coding

Table 7-2 Alternative types of processing in expanded lossy DCT-based mode.

7.5.4 Lossless Mode

The lossless mode shown in Figure 7-17 uses single pixels as data units during image preparation. Between 2 and 16 bits can be used per pixel. Although all pixels of an image must use the same precision, one can also conceive of adaptive pixel precision.

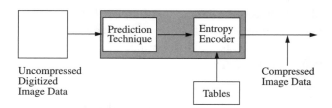

Figure 7-17 Lossless mode, based on prediction.

In this mode, image processing and quantization use a predictive technique instead of transformation coding. For each pixel X as shown in Figure 7-18, one of eight possible predictors is selected. The selection criterion is the best possible prediction of the value of X from the already known adjacent samples A, B, and C. Table 7-3 lists the specified predictors.

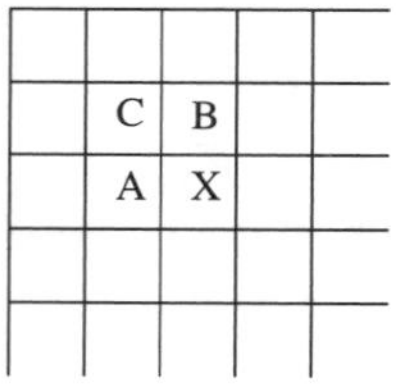

Figure 7-18 Basis of prediction in lossless mode.

The number of the chosen predictor, as well as the difference between the prediction and the actual value, are passed to the subsequent entropy encoding, which can use either Huffman or arithmetic coding.

In summary, this mode supports two types of processing, each with between 2 and 16 bits per pixel. Each variant can use either Huffman coding or arithmetic coding.

Selection Value	Prediction
0	No prediction
1	X = A
2	X = B
3	X = C
4	X = A + B + C
5	X = A + (B - C)/2
6	X = B + (A - C)/2
7	X = (A + B)/2

Table 7-3 Predictors in lossless mode.

7.5.5 Hierarchical Mode

The hierarchical mode can either use one of the lossy DCT-based algorithms described above or alternatively use the lossless compression technique, as the need arises. The main feature of this mode is the encoding of an image at different resolutions, that is, the compressed image contains images at several resolutions. To do this, the prepared digital image is first reduced by a factor of 2^n and compressed. The original image is then reduced by a factor of 2^{n-1} vertically and horizontally. The previously compressed image is subtracted from this, and the result is once again compressed. This process is successively repeated until the full resolution of the image is compressed.

Hierarchical coding is computationally intensive and requires considerable storage space. The advantage is the compressed image is available at different resolutions. Applications working with lower resolutions thus do not need to first decode the whole image and then reduce the resolution. In other words, scaling becomes cheap. According to the authors' practical experience, it is often more efficient to display an image in its full resolution than to first scale it down and display a smaller image. Yet, in the case of images encoded according to the hierarchical JPEG mode, the display of a reduced size picture requires less time to process than one of higher resolution.

7.6 H.261 (*p*×64) and H.263

The driving force behind the H.261 video standard was and is ISDN. One or both B-channels of a narrowband ISDN connection can transfer video data, for example, in addition to speech. This requires that both partners connected via the channel have to use the same coding of video data. In a narrowband ISDN connection, exactly two B-channels and one D-channel are available at the user interface. The European ISDN hierarchy also allows a connection with 30 B-channels, intended for PABX. In the following, when we speak of *the* B-channel, this refers to one or more channels. Early on, the primary applications of ISDN were videophone and videoconferencing systems. These dialogue applications require that coding and decoding be carried out in real time. In 1984, Study Group XV of CCITT established a committee of experts to draw up a standard for the compression of moving pictures [Lio91].

The resulting CCITT Recommendation H.261 *Video CoDec for Audiovisual Services at p×64Kbit/s* [ITUC90] was finalized after five years of work and accepted in December 1990. In this context, codec (Coder/Decoder) refers to encoding and decoding, or to compression and decompression. North America adopted the recommendation with slight modifications. Since data rates of $p \times 64$ Kbit/s are considered, the recommendation is also known as $p \times 64$.

The ITU Study Group XV Recommendation H.261 was developed for real-time processing during encoding and decoding. The maximum combined signal delay for compression and decompression must not exceed 150 ms. If the end-to-end delay is too great, the subjective interactivity of a dialogue application using this standard suffers considerably.

H.263 is a provisional ITU-T Standard published in 1996 to replace H.261 for many applications. H.263 was designed for low bit rate transmission. Early designs called for data rates under 64 Kbit/s, though this was later revised. As part of the ITU-T H.320 family of standards (recommendation for real-time voice, data, and video over V.34 modems using a conventional GSTN telephone network), H.263 is used for a wide range of bit rates (not just low bit rate applications).

With respect to efficiency of video compression, H.263 is one of the best techniques available today. The H.263 coding algorithm is similar to that of H.261, with some improvements and changes to further improve performance and add error correction.

The following are the key differences between the H.261 and H.263 coding algorithms:

- H.263 uses half pixel precision for motion compensation, while H.261 uses full pixel precision with a "loop filter."
- Some parts of the hierarchical structure of the data stream are optional in H.263, so that the codec can be configured for a lower bit rate or better error correction.
- H.263 includes four optional, negotiable parameters to improve performance:
    - the unrestricted motion vector mode,
    - the syntax-based arithmetic coding mode,
    - the advanced prediction mode,
    - and forward and backward frame prediction (similar to the P and B frames in MPEG).

- H.263 can often achieve the same quality as H.261 with less than half as many bits by using the improved negotiable options.
- H.263 supports five resolutions. In addition to QCIF and CIF, which H.261 supports, H.263 also supports SQCIF, 4CIF, and 16CIF. SQCIF has about half the resolution of QCIF. 4CIF and 16CIF correspond to four and 16 times the resolution of CIF, respectively. Support for 4CIF and 16CIF means that the codec can unquestionably vie with other high bit rate coding standards, such as MPEG.

7.6.1 Image Preparation

Unlike JPEG, H.261 defines a very precise image format. The image refresh frequency at the input must be $30000/1001 = 29.97$ frames/s. During encoding, it is possible to generate a compressed image sequence with a lower frame rate of, for example, 10 or 15 frames per second. Images cannot be presented at the input to the coder using interlaced scanning. The image is encoded as a luminance signal (Y) and chrominance difference signals C_b, C_r, according to the CCIR 601 subsampling scheme (2:1:1), which was later adopted by MPEG.

Two resolution formats, each with an aspect ratio of 4:3 are defined. The so-called Common Intermediate Format (CIF) defines a luminance component of 352 lines, each with 288 pixels. As per the 2:1:1 requirement, the chrominance components are subsampled with 176 lines, each with 144 pixels.

Quarter CIF (QCIF) has exactly half the resolution in all components (i.e., 176×144 pixels for the luminance and 88×72 pixels for the other components). All H.261 implementations must be able to encode and decode QCIF. CIF is optional.

The following example illustrates the compression ratio necessary to encode even an image with the low resolution of QCIF for the bandwidth of an ISDN B-channel. At 29.97 frames/s, the uncompressed QCIF data stream has a data rate of 9.115 Mbit/s. At the same frame rate, CIF has an uncompressed data rate of 36.45 Mbit/s. The image to be processed should be compressed at a rate of ten frames per second. This leads to a necessary compression ratio for QCIF of about 1:47.5, which can be easily supported by today's technology.

For CIF, a corresponding reduction to about six ISDN B-channels is possible. H.261 divides the Y as well as the C_b and C_r components into blocks of 8×8 pixels. A macro block is the result of combining four blocks of the Y matrix with one block each from the C_b and C_r components. A group of blocks consists of 3×11 macro blocks. A QCIF image thus consists of three groups of blocks and a CIF image consists of twelve groups of blocks.

7.6.2 Coding Algorithms

The H.261 standard uses two different methods of coding: intraframe and interframe. Intraframe coding under H.261 considers only the data from the image being coded; this corresponds to *intrapicture* coding in MPEG (see Section 7.7.1). Interframe coding in H.261 uses data from other images and corresponds to P-frame coding in MPEG (see Section 7.7.1). The H.261 standard does not prescribe any criteria for using one mode or the other depending on specific parameters. The decision must be made during encoding and thus depends on the specific implementation.

H.263, unlike H.261, recommends four negotiable modes of intraframe coding. These can be used separately or together. An exception is that the advanced prediction mode requires use of the unrestricted motion vector mode.

The new intraframe coding modes added to H.263 are briefly described below:

1. Syntax-based arithmetic coding mode
   defines the use of arithmetic coding instead of variable-length coding. This results in identical image recoverability with better compression efficiency.
2. PB-frame mode
   can increase the frame rate without changing the bit rate by coding two images as one unit. The images must be a predicted, or P frame, and a B frame (as defined by MPEG; see Section 7.7.1), which is predicted bidirectionally from the preceding P frame and the current P frame.
3. Unrestricted motion vector mode
   makes it possible for motion vectors to point outside image boundaries. This is particularly useful for small images with movements in the direction of the edges.
4. Advanced prediction mode
   uses the Overlapped Block Motion Compensation (OBMC) technique for P-frame luminance. The coder can use one 16×16 vector or four 8×8 vectors for each macro block. Using smaller vectors requires more bits but yields better predictions, and in particular, fewer artifacts.

Like JPEG, for intraframe coding, each block of 8×8 pixels is transformed into 64 coefficients using DCT. Here also, DC-coefficients are quantized differently than AC-coefficients. The next step is to perform entropy encoding using variable-length code words.

In interframe coding, a prediction method is used to find the most similar macro block in the preceding image. The motion vector is defined by the relative position of the previous macro block with respect to the current macro block. According to H.261, a coder does not need to be able to determine a motion vector. Thus a simple H.261 implementation can always consider only differences between macro blocks located at the same position of successive images. In this case, the motion vector is always the zero vector. Next, the motion vector and the DPCM-coded macro block are processed. The latter is transformed by the DCT if and only if its value exceeds a certain threshold value. If the difference is less than the threshold, then the macro block is not encoded any further and only the motion vector is processed. The components of the motion vector are entropy encoded using variable-length coding. This is lossless.

Transformed coefficients are all quantized linearly and entropy encoded using variable-length code words.

Optionally, an optical low pass filter can be inserted between the DCT and the entropy encoding. This filter deletes any remaining high-frequency noise. H.261 implementations need not incorporate this filter.

H.261 uses linear quantization. The step size is adjusted according to the amount of data in a transmission buffer, thereby enforcing a constant data rate at the output of the coder. This feedback also influences the image quality.

7.6.3 Data Stream

According to H.261/H.263, a data stream is divided into several layers, the bottom layer containing the compressed images. Some interesting properties of H.261 and H.263 are mentioned below (for further details, see [ITUC90]):

- The data stream contains information for error correction, although use of external error correction, e.g. H.223, is recommended.
- Each image in H.261 includes a five-bit number that can be used as a temporal reference. H.263 uses eight-bit image numbers.
- During decoding, a command can be sent to the decoder to "freeze" the last video frame displayed as a still frame. This allows the application to stop/freeze and start/play a video scene without any additional effort.
- Using an additional command sent by the coder, it is possible to switch between still images and moving images. Alternatively, instead of using explicit commands, a time out signal can be used.

7.6.4 H.263+ and H.263L

H.263+ is a planned extension of the existing H.263 standard. Improvements will probably be rather small, especially in the area of coding options. Examples of methods that will be incorporated in H.263+ are the 4×4 DCT, improved intra coding, and a deblocking filter in the prediction loop.

H.263L is further improvement on H.263 with a longer time horizon than H.263+. Here greater changes are expected. H.263L could coincide with the development of MPEG-4.

7.7 MPEG

MPEG was developed and defined by ISO/IEC JTC1/SC 29/WG 11 to cover motion video as well as audio coding. In light of the state of the art in CD technology, the goal was a compressed stream data rate of about 1.2Mbit/s. MPEG specifies a maximum data rate of 1,856,000 bit/s, which should not be exceeded [ISO93b]. The data rate for each audio channel can be chosen between 32 and 448 Kbit/s in increments of 16 Kbit/s. This data rate enables video and audio compression of acceptable quality. Since 1993, MPEG has been an International Standard (IS) [ISO93b]. MPEG explicitly takes into account developments in other standardization activities:

- JPEG: Since a video sequence can be regarded as a sequence of still images and the JPEG standard development was always ahead of the MPEG standardization, the MPEG standard makes use of the results of the JPEG work.
- H.261: Since the H.261 standard already existed during the MPEG work, the MPEG group strived to achieve at least a certain compatibility between the two standards in some areas. This should simplify implementations of MPEG that also support H.261. In any case, technically MPEG is the more advanced technique. Conversely, H.263 borrowed techniques from MPEG.

Although mainly designed for asymmetric compression, a suitable MPEG implementation can also meet symmetric compression requirements. Asymmetric coding requires considerably more effort for encoding than for decoding. Compression is carried out once, whereas decompression is performed many times. A typical application area is retrieval systems. Symmetric compression is characterized by a comparable effort for the compression and decompression processing. This is a requirement for interactive dialogue applications, as is a bounded processing delay for the processing.

Besides the specification of video coding [Le 91, VG91] and audio coding, the MPEG standard provides a system definition, which describes the combination of individual data streams into a common stream.

7.7.1 Video Encoding

In the image preparation phase (according to the reference scheme shown in Figure 7-1), MPEG, unlike JPEG but similar to H.263, defines the format of an image very precisely.

7.7.1.1 Image Preparation

An image must consist of three components. In addition to the luminance Y there are two color difference signals C_r and C_b (similar to the YUV format). The luminance component has twice as many samples in the horizontal and vertical axes as the other components, that is, there is color subsampling. The resolution of the luminance component should not exceed 768×576 pixels. The pixel depth is eight bits in each component.

An MPEG data stream also contains information that is not part of a data stream compressed according to the JPEG standard, for example the pixel aspect ratio. MPEG supports 14 different pixel aspect ratios. The most important are:

- A square pixel (1:1) is suitable for most computer graphics systems.
- For a 625-line image, a ratio of 16:9 is defined (European HDTV).
- For a 525-line image, a ratio of 16:9 is also defined (U.S. HDTV).
- For 702×575 pixel images, an aspect ratio of 4:3 is defined.
- For 711×487 pixel images, an aspect ratio of 4:3 is also defined.

The image refresh frequency is also encoded in the data stream. So far, eight frequencies have been defined (23.976 Hz, 24 Hz, 25 Hz, 29.97 Hz, 30 Hz, 50 Hz, 59.94 Hz, and 60 Hz), so no low image refresh frequencies are permitted.

Temporal prediction of still images usually yields a considerable data reduction. Areas within an image with strong, irregular motions can only be reduced by a ratio similar to that of intraframe coding. The use of temporal predictors requires the storage of a huge amount of previously determined information and image data. The extent to which prediction is employed can be determined by balancing the required storage capacity against the achievable compression rate.

In most cases, predictive coding only makes sense for parts of an image and not for the whole image. The image is thus divided into areas called macro blocks. An MPEG macro block is partitioned into 16×16 pixels for the luminance component and 8×8 pixels for each of the two chrominance components. These sizes are well suited for compression based on motion estimation. This is a compromise between the computational effort required for estimation and the resulting data reduction. A macro block is formed of six blocks of 8×8 pixels, ordered as follows: first four blocks for the luminance component then the two chrominance blocks. There are no user-defined MCUs as in JPEG since, given the defined frame rates, the maximum time to present an image is 41.7 ms. The three components are compressed and decompressed together. From the MPEG user's perspective, there is no fundamental advantage to progressive image display over sequential display.

7.7.1.2 Image Processing

For the image processing stage, MPEG supports four types of image coding. The reason for this is the contradictory demands of efficient coding and random access. In order to achieve a high compression ratio, temporal redundancies of successive images need to be exploited. Fast random access requires that images be coded individually. Hence the following image types are distinguished:

- I frames (intra coded pictures) are coded without using information about other frames (intraframe coding). An I frame is treated as a still image. Here MPEG falls back on the results of JPEG. Unlike JPEG, real-time compression must be possible. The compression rate is thus the lowest within MPEG. I frames form the anchors for random access.
- P frames (predictive coded pictures) require information about previous I and/or P frames for encoding and decoding. Decoding a P frame requires decompression of

the last I frame and any intervening P frames. In return, the compression ratio is considerably higher than for I frames. A P frame allows the following P frame to be accessed if there are no intervening I frames.

- B frames (bidirectionally predictive coded pictures) require information from previous and following I and/or P frames. B frames yield the highest compression ratio attainable in MPEG. A B frame is defined as the difference from a prediction based on a previous and a following I or P frame. It cannot, however, ever serve as a reference for prediction coding of other pictures.
- D frames (DC coded pictures) are intraframe-coded and can be used for efficient fast forward. During the DCT, only the DC-coefficients are coded; the AC-coefficients are ignored.

Figure 7-19 shows a sequence of I, P, and B frames. This example illustrates the prediction for the first P frame and the bidirectional prediction for a B frame. Note that the order in which the images are presented differs from the actual decoding order if B frames are present in an MPEG-coded video stream.

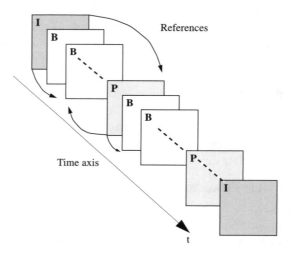

Figure 7-19 Types of individual images in MPEG: I, B, and P frames.

The pattern of I, P, and B frames in a sequence is determined by the MPEG application. For random access, the ultimate resolution would be attained by encoding the entire stream using I frames. The highest compression rate can be achieved by using as many B frames as possible. For practical applications, the sequence IBBPBBPBB IBBPBBPBB... has proven to be useful. This permits random access with a resolution of nine still images (i.e., about 330ms) and still provides a very good compression ratio. Every 15 images includes one I frame.

The following detailed description of image processing, quantization and entropy encoding distinguishes the four image types.

7.7.1.3   I Frames

I frames are encoded by performing a DCT on the 8×8 blocks defined within the macro blocks, as in JPEG. The DC-coefficients are then DPCM coded, and differences between consecutive blocks of each component are calculated and transformed into variable-length code words. AC-coefficients are run-length encoded and then transformed into variable-length code words. MPEG distinguishes two types of macro blocks: those that contain only coded data and those that additionally contain a parameter used for scaling the characteristic curve used for subsequent quantization.

7.7.1.4   P Frames

The coding of P frames exploits the fact that in consecutive images, some areas of the image often shift (move), but do not change. To encode a block, the most similar macro block in the preceding image must be determined. This is done by calculating the differences between the absolute values of the luminance component for each macro block in the preceding image. The macro block with the lowest sum of differences is the most similar. MPEG does not specify an algorithm for motion estimation, but rather specifies the coding of the result. Only the motion vector (the spatial difference between the two macro blocks) and the small difference between the macro blocks need to be encoded. The search range, that is, the maximum length of the motion vector, is not defined by the standard. As the search range is increased, the motion estimation becomes better, although the computation becomes slower [ISO93b].

P frames can consist of macro blocks as in I frames, as well as six different predictive macro blocks. The coder essentially decides whether to code a macro block predictively or like an I frame and whether it should be coded with a motion vector. A P frame can thus also include macro blocks that are coded in the same way as I frames.

In coding P-frame-specific macro blocks, differences between macro blocks as well as the motion vector need to be considered. The difference values between all six 8×8 pixel blocks of a macro block being coded and the best matching macro block are transformed using a two-dimensional DCT. Further data reduction is achieved by not further processing blocks where all DCT coefficients are zero. This is coded by inserting a six-bit value into the encoded data stream. Otherwise, the DC- and AC-coefficients are then encoded using the same technique. This differs from JPEG and from the coding of I frame macro blocks. Next, run-length encoding is applied and a variable-length coding is determined according to an algorithm similar to Huffman. Since motion vectors of adjacent macro blocks often differ only slightly, they are DPCM coded. The result is again transformed using a table; a subsequent calculation performs a transformation into variable-length coded words.

7.7.1.5 B Frames

In addition to the previous P or I frame, B-frame prediction takes into account the following P or I frame. The following example illustrates the advantages of bidirectional prediction:

In a video sequence, a ball moves from left to right in front of a static background. In the left area of the scene, parts of the image appear that in previous images were covered by the moving ball. A prediction of these areas would ideally be derived from the following, not from the previous, image. A macro block can be derived from macro blocks of previous and following P and/or I frames. Motion vectors can also point in orthogonal directions (i.e., in x direction and in y direction). Moreover, a prediction can interpolate two similar macro blocks. In this case, two motion vectors are encoded and one difference block is determined between the macro block to be encoded and the interpolated macro block. Subsequent quantization and entropy encoding are performed as for P-frame-specific macro blocks. Since B frames cannot serve as reference frames for subsequent decoding, they need not be stored in the decoder.

7.7.1.6 D Frames

D frames contain only the low-frequency components of an image. A D-frame always consists of one type of macro block and only the DC-coefficients of the DCT are coded. D frames are used for fast-forward display. This could also be realized by a suitable placement of I frames. For fast-forward, I frames must appear periodically in the data stream. In MPEG, slow-rewind playback requires considerable storage. All images in a so-called group of pictures must decoded forwards and stored before they can be displayed in reverse.

7.7.1.7 Quantization

Concerning quantization, it should be noted that AC-coefficients of B and P frames are usually very large values, whereas those of I frames are very small. Thus, MPEG quantization quantization adjusts itself accordingly. If the data rate increases too much, quantization becomes more coarse. If the data rate falls, then quantization is performed with finer granularity.

7.7.2 Audio Coding

MPEG audio coding is compatible with the coding of audio data used for Compact Disc Digital Audio (CD-DA) and Digital Audio Tape (DAT). The most important criterion is the choice of sample rate of 44.1 kHz or 48 kHz (additionally 32 kHz) at 16 bits per sample value. Each audio signal is compressed to either 64, 96, 128, or 192 Kbit/s.

Three quality levels (layers) are defined with different encoding and decoding complexity. An implementation of a higher layer must be able to decode the MPEG audio signals of lower layers [Mus90].

Similar to two-dimensional DCT for video, a transformation into the frequency domain is applied for audio. The Fast Fourier Transform (FFT) is a suitable technique. As shown in Figure 7-20, the relevant portion of the spectrum is divided into 32 non-overlapping subbands. The audio signal is thus split into 32 subbands. Different components of the spectrum can then be quantized differently. In parallel with the actual FFT, the noise level in each subband is determined using a psychoacoustic model. At a higher noise level, a coarser quantization is performed. A lower noise level results in finer quantization. In the first and second layers, the appropriately quantized spectral components are simply PCM-encoded. The third layer additionally performs Huffman coding.

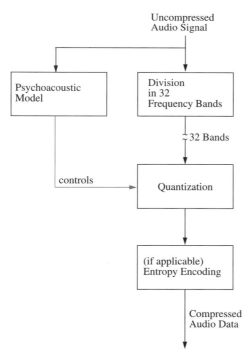

Figure 7-20 MPEG audio encoding.

Audio coding can be performed on stereo sound, a single channel, or two independent channels. MPEG provides for two types of stereo sound. In the first case, two channels are processed completely independently. In the joint stereo mode, MPEG achieves a higher compression ratio by exploiting redundancies between the two channels.

Each layer defines 14 fixed bit rates for the coded audio data stream, which in MPEG are addressed by a bit rate index. The minimal value is always 32 Kbit/s. The layers support different maximal bit rates: layer 1 allows for a maximum of 448 Kbit/s, layer 2 for 384 Kbit/s, and layer 3 for 320 Kbit/s. For layers 1 and 2, a decoder is not required to support a variable bit rate. In layer 3, a variable bit rate is specified by allowing the bit rate index to be switched. In layer 2, not all combinations of bit rate and mode are allowed:

- 32 Kbit/s, 48 Kbit/s, 56 Kbit/s, and 80 Kbit/s are only allowed for a single channel.
- 64 Kbit/s, 96 Kbit/s, 112 Kbit/s, 128 Kbit/s, 160 Kbit/s, and 192 Kbit/s are allowed for all modes.
- 224 Kbit/s, 256 Kbit/s, 320 Kbit/s, and 384 Kbit/s are allowed for the stereo, joint stereo and dual channel modes.

7.7.3 Data Stream

Like JPEG, MPEG specifies a precise syntax for the compressed audio and video data streams.

7.7.3.1 Audio Stream

An audio stream is comprised of frames, which are made up of audio access units, which in turn are divided into slots. At the lowest coding complexity (layer 1), a slot consists of 4 bytes; otherwise it is one byte. Frames consist of a fixed number of samples. An audio access unit is the smallest compressed audio sequence that can be completely decoded independently of all other data. At 48 kHz, the audio access units contained in a frame have a play time of 8 ms; at 44.1 kHz, the play time is 8.7 ms; and at 32 kHz, the play time is 12 ms. In the case of stereo, data from both channels are included in one frame.

7.7.3.2 Video Stream

A video stream is comprised of 6 layers:

1. At the highest level, the sequence layer, data buffering is handled. For example, it does not make sense to generate a data stream that places excessive demands on storage space in the decoder. Among other things, the beginning of the sequence layer thus includes two entries: the constant bit rate of the sequence and the minimum storage capacity required during decoding.

   A video buffer verifier is added after the quantizer that uses the bit rate resulting from decoding in order to monitor the decoding delay. The video buffer verifier influences the quantizer and forms a type of control loop. Successive sequences can have varying bit rates. During decoding of several consecutive sequences, often no data are output between the end of one sequence and the beginning of the next sequence because the underlying decoder parameters need to be updated and an initialization performed.

2. The group of pictures layer is the next layer. This layer contains at least an I frame, which must be one of the first images. Random access to this image is always possible. At this layer, it is possible to distinguish between the order of images in the data stream and their display order. The data stream must always begin with an I frame so that the decoder can first decode and store the reference image. In order of presentation, however, B frames can precede the I frame. The following example illustrates the difference between decoding order and display order:

Decoding order													
Type of Frame	I	P	B	B	P	B	B	P	B	B	I	B	B
Frame number	1	4	2	3	7	5	6	10	8	9	13	11	12

Display order													
Type of Frame	I	P	B	B	P	B	B	P	B	B	I	B	B
Frame number	1	2	3	4	5	6	7	8	9	10	11	12	13

3. The picture layer contains a whole still image. The image's temporal reference is defined using an image number. This number is shown below each image in the example above. There are also data fields defined in this layer that are not yet used in MPEG. These fields are reserved for future extensions and may not be used by the decoder.
4. The next layer is the slice layer. Each slice consists of macro blocks, the number of which can vary from image to image. A slice also includes the scaling used for DCT quantization of all its macro blocks.
5. The next layer is the macro block layer. This contains a macro block with the properties described above.
6. The lowest layer is the block layer, also described above.

7.7.3.3 System Definition

The MPEG system definition also specifies the combination of both data streams into a single stream. The most important task of this process is the actual multiplexing. It includes the coordination of input data streams with output data streams, clock adjustment, and buffer management. Data streams as defined by ISO 11172 are divided into separate packs. The decoder gets the information it needs from the multiplexed data stream. For example, the maximal data rate is included at the beginning of every ISO 11172 data stream in the first pack.

This definition of the data stream makes the implicit assumption that it is possible to read such a header prior to the first (possibly random) data access from a secondary storage medium. In dialogue services over communications networks, such as telephone or videophone, a participant will always get the header information first. On the other hand, this would be inconvenient in a conferencing application where a user can join at any time. The required header information would not then be immediately available, since it transmitted only at the beginning of an ISO 11172 data stream. One could, however, define a protocol to supply the header upon request.

MPEG also supplies time stamps necessary for synchronization in ISO 11172 data streams. These concern the relationship between multiplexed data streams within an ISO 11172 data stream and not other ISO 11172 data streams that may exist.

It should be noted that MPEG does not prescribe compression in real-time. Moreover, MPEG defines the decoding process but not the decoder itself.

7.7.4 MPEG-2

Today one assumes that the quality of a video sequence compressed according to the MPEG standard at the maximum data rate of about 1.5 Mbit/s cannot be substantially improved. Here only the results (the compression ratio and the quality) should count, not the required processing expenditure. Thus a video compression technique is being developed for rates up to 100 Mbit/s. This is known as "MPEG-2" [ISO93a]. The previously established technique is now known as "MPEG-1," while MPEG-2 aims at a higher image resolution, similar to the CCIR 601 digital video studio standard.

To ensure a harmonized solution that covered a wide range of applications, the ISO/IEC Working Group ISO/IEC JTC1/SC29/WG11 developed MPEG-2 in close cooperation with the ITU-TS Study Group 15 Experts Group for ATM Video Coding. In addition to these two groups, other representatives of the ITU-TS, EBU, ITU-RS, SMPTE, and North American HDTV developers also worked on MPEG-2.

The MPEG group developed the MPEG-2 Video standard, which specifies the coded bit stream for higher quality digital video. As a compatible extension, MPEG-2 video builds on the MPEG-1 standard by supporting interlaced video formats and a number of other advanced features, including support for HDTV.

	Profile name	Simple Profile	Main Profile	SNR Scaling Profile	Spatial Scaling Profile	High Profile
Profiles	**Characteristics of profile**	no B frames	B frames			
		4:2:0				4:2:0 or 4:2:2
		not scalable		SNR scalable	SNR or spatially scalable	
Levels	**High Level** 1920 pixels/line 1152 lines		≤ 80 Mbit/s			≤ 100 Mbit/s
	High-1440 Level 1440 pixels/line 1152 lines		≤ 60 Mbit/s		≤ 60 Mbit/s	≤ 80 Mbit/s
	Main Level 720 pixels/line 572 lines	≤ 15 Mbit/s	≤ 15 Mbit/s	≤ 15 Mbit/s		≤ 20 Mbit/s
	Low Level 352 pixels/line 288 lines		≤ 15 Mbit/s	≤ 15 Mbit/s		

Table 7-4 MPEG-2 Profiles and Levels. (Empty cells contain undefined values.)

As a generic international standard, MPEG-2 Video was defined in terms of extensible profiles, each of which supports the features needed by a class of applications. The MPEG-2 Main Profile was defined to support digital video transmission in the range of about 2 to 80 Mbit/s using cable, satellite, or other broadcast channels, as well as to support digital storage and other communications applications. The parameters of the Main Profile and the High Profile are suitable for supporting HDTV.

The MPEG experts also extended the features of the Main Profile by defining a hierarchical/scalable profile. This profile aims to support applications such as compatible terrestrial TV, packet-network video systems, backwards compatibility with existing standards (MPEG-1 and H.261), and other applications requiring multi-level coding. For example, such a system could offer a consumer the option of using a small portable receiver to decode standard definition TV from a broadcast signal, or using a larger fixed receiver to decode HDTV from the same signal.

The MPEG-2 profiles arranged in a 5×4 matrix as shown in Table 7-4. The degree of functionality increases along the horizontal axis. The vertical axis indicates levels with increased parameters, such as smaller and larger frame sizes. For example, the Main Profile in the Low Level specifies 352 pixels/line with 288 lines/frame and 30 frames/s, without any B frames and a data rate not to exceed 4 Mbit/s. The Main Profile in the High Level specifies 1920 pixels/line with 1152 lines/frame and 60 frames/s with a data rate not to exceed 80 Mbit/s.

MPEG-2 incorporates, in a way similar to the hierarchical mode of JPEG, scaling of compressed video [GV92]. In particular, video is compressed at different qualities during encoding, so that different alternatives are available during decompression [Lip91, GV92]. Scaling may act on various parameters:

- Spatial scaling facilitates decompression of image sequences with different horizontal and vertical resolutions. For example, a single data stream could include images with 352×288 pixels (H.261 CIF format), 360×240 pixels, 704×576 pixels (a format according to CCIR 601) and, for example 1,250 lines at an aspect ratio of 16:9. These resolutions refer to the luminance component; the chrominance component is subsampled by a factor of two. This can be implemented using a pyramid for the level of the DCT-coefficients [GV92], that is, 8×8 DCT, 7×7 DCT, 6×6 DCT, and other transformations can be formed. From a technical standpoint, only steps by a factor of two are useful.

- Scaling the data rate allows for playback at different frame rates. In MPEG-1, this functionality is achieved by using D frames. This can be implemented in MPEG-2 by using I frames, given a suitable distribution of I frames within the data stream. This condition must be met not only for a group of pictures, but also for the entire video sequence.

- Amplitude scaling can be interpreted as affecting either the pixel depth or the resolution at which DCT-coefficients are quantized. This then leads to a layered coding and to the possibility of progressive image presentation, which is not relevant for video data. However, this can be of interest if certain images of an image sequence can be extracted from the data stream as still images.

Scaling is one of the most important MPEG-2 extensions to MPEG-1. MPEG-2 also, for example, minimizes the effects of the loss of individual ATM cells in an MPEG-2 coded data stream. Sequences of different frame types (I, P, B) should be defined to minimize end-to-end delay at a given data rate.

The MPEG-2 group developed the MPEG-2 Audio Standard for low bit rate coding of multichannel audio. MPEG-2 supplies up to five full bandwidth channels (left, right, center, and two surround channels), plus an additional low-frequency enhancement channel and/or up to seven commentary/multilingual channels. The MPEG-2 Audio Standard also extends MPEG-1 coding of stereo and mono channels with half sampling rates (16 kHz, 22.05 kHz, and 24 kHz), which significantly improves quality at 64 Kbit/s or less per channel.

The MPEG-2 Audio Multichannel Coding Standard is backwards compatible with the existing MPEG-1 Audio Standard. The MPEG group carried out formal subjective testing of the proposed MPEG-2 Multichannel Audio codecs and up to three nonbackward-compatible codecs working at rates from 256 to 448 Kbit/s.

7.7.4.1 MPEG-2 System

MPEG-2 addresses video together with associated audio. Note that in order to provide a precise description, this section uses terminology from the original MPEG-2 specification. MPEG-2 defines how an MPEG-2 system can combine audio, video, and other data in a single stream, or in multiple streams, suitable for storage and transmission. MPEG-2 imposes syntactic and semantic rules that are necessary and sufficient to synchronize the decoding and presentation of video and audio information, while ensuring that the decoder's coded data buffers do not overflow or run out of data. The streams include time stamps used in the decoding, presentation, and delivery of the data.

In the first step, the basic multiplexing approach adds system-level information to each stream. Each stream is then divided into packets, forming a Packetized Elementary Stream (PES). Next, the PESs are combined into a Program Stream or a Transport Stream. Both streams were developed to support a large number of known or anticipated applications. They thus incorporate a significant degree of flexibility while ensuring interoperability between different device implementations.

- The Program Stream resembles the MPEG-1 stream, but should only be used in a relatively error-free environment. Program stream packets are of variable length. Timing information in this stream can be used to implement a constant end-to-end delay (along the path from the input of the coder to the output of the decoder).
- The Transport Stream bundles the PESs and one or more independent time bases into a single stream. This stream was developed for use with lossy or noisy media. Each packet has a length of 188 bytes, including a four-byte header. The Transport Stream is well-suited for transmission of digital television and video telephony over fiber, satellite, cable, ISDN, ATM, and other networks, as well as for storage on digital video tape and other devices.

A conversion between the Program Stream and the Transport Stream is possible and sensible. Note that the MPEG-2 buffer management specification limits the end-to-end delay for audio and video data to under one second, a value unacceptably high for users of dialogue mode applications.

A typical MPEG-2 video stream has a variable bit rate. By using a video buffer as specified in the standard, a constant bit rate can be enforced, at the cost of varying quality.

MPEG-2 reached the CD (Committee Draft) status in late 1993 and required three additional months to become a DIS (Draft International Standard). After a six-month ballot period, the DIS became an IS (International Standard). Originally, there were plans to specify an MPEG-3 standard covering HDTV. However, during the development of MPEG-2, it was found that scaling up could easily meet the requirements of HDTV. Consequently, MPEG-3 was dropped.

7.7.5 MPEG-4

Work on another MPEG initiative for very low bit rate coding of audio visual programs started in September 1993 in the ISO/IEC JTC1. Formally designated ISO/IEC 14496, MPEG-4 was published in November 1998 and adopted as an international standard in January 1999.

MPEG-4 incorporates new algorithmic techniques, such as model-based image coding of human interaction with multimedia environments and low bit-rate speech coding for use in environments like the European Mobile Telephony System (GSM). The most important innovation is improved flexibility. Developers can use the compression method in different ways and configure systems to support a multitude of applications. MPEG-4 is thus not a fixed standard suitable only for a few applications. Moreover, MPEG-4 integrates a large number of audiovisual data types, for example natural and synthetic, aimed at a representation in which content-based interactivity is supported for all media types. MPEG-4 thus allows the design of audiovisual systems that are oriented around specific users yet are compatible with other systems.

Most currently available audiovisual services allow only playback functionality. In contrast, MPEG-4 places a strong emphasis on interactivity. Support for random access to audio and video scenes as well as the ability to revise content are thus among the main objectives of the MPEG-4 standardization. MPEG-4 provides for a universal coding of various forms of audiovisual data, called audiovisual objects. In other words, MPEG-4 seeks to represent the real world as a composition of audiovisual objects. A script then describes the spatial and temporal relationships among the objects. Using this form of representation, the user can interact with the various audiovisual objects in a scene in a manner that corresponds to the actions of everyday life.

Although this content-based approach to scene representation may seem obvious to the human user, it actually represents a revolution in video representation since it enables a quantum leap in the functionality that can be offered to the user. If a scene is represented as a collection of a few or many independent audiovisual objects, the user has the opportunity to play with the contents of the scene, for example by changing the properties of some objects (e.g., position, motion, texture, or form), accessing only selected parts of a scene, or even using cut and paste to insert objects from one scene into another scene. Interaction with contents is thus a central concept of MPEG-4.

Another major shortcoming of other audiovisual coding standards is the restricted number of audio and video data types used. MPEG-4 attempts to smoothly integrate natural and synthetic audiovisual objects, for example mono, stereo, and multiple channel audio. Moreover, MPEG-4 supports either 2-D or 3-D (mono or stereo, respectively) video modes, as well as these from additional camera perspectives (so-called multiview video). This integration should be extended with regard to audiovisual interrelationships so that processing can incorporate the mutual influence and dependence between different types of information. Moreover, new and already available analysis and coding tools are being integrated into MPEG-4. In the future, MPEG-4 as a whole will be further developed as new or better tools, data types, or functionality becomes available.

The quick pace of technological progress in recent years has accentuated the inflexibility of standards that do not allow for the continuous development of hardware and methods. Such standards implement only specific solutions that very quickly no longer reflect technological developments. Flexibility and extensibility are thus important objectives of MPEG-4. MPEG-4 provides flexibility and extensibility through the "MPEG-4 Syntactic Description Language (MSDL)." The MSDL approach to extending audiovisual coding standards is revolutionary. Not only can new algorithms be integrated by selecting and integrating predefined tools (level 1), but they can be "learned" in that the coder can download new tools. At the same time, even MSDL is subject to further development since the MPEG-4 standard can always incorporate new tools, algorithms, and concepts in MSDL to support new or improved functionalities.

The areas of telecommunications, computers, and television/film are converging and are in the position of being able to exchange elements that previously were typical for only one of the three areas. This convergence is more likely viewed as evolutionary because it is taking place through gradual transitions. Because qualitatively new multimedia services are emerging from this convergence, a logical consequence is that new requirements will be placed on coding and transmission techniques. With its official focus on the three driving forces of content and interaction, integration and flexibility and extensibility, MPEG-4 seems to be the appropriate answer to such requirements.

7.7.5.1 MPEG-4 Extensions Compared with MPEG-2

The vision of the MPEG-4 standard can be best explained using the eight new or improved functionalities identified in the MPEG-4 Proposal Package Description. These were the result of a meeting held to determine which features could be of significance in the near future but that were not (or only partially) supported by present-day coding standards.

MPEG-4 must additionally provide some other important so-called standard functions included in previously available standards, for example synchronization of audio and video, low-delay modes, and interoperability between networks. Unlike the new or improved functions, the standard functions can be provided by standards that either exist or are under development.

MPEG-4 makes it possible to scale content, spatial and temporal resolution, quality, and complexity with finer granularity. This content-based scalability implies the existence of a priority mechanism for a scene's objects. The combination of various scaling methods can lead to interesting scene representations, whereby more important objects are represented using higher spatial-temporal resolution. Content-based scaling is a key part of the MPEG-4 vision, since other features can easily be implemented once a list of a scene's important objects is available. In some cases, these and related features may require analyzing a scene to extract the audiovisual objects, depending on the application and on previous availability of composition information.

MPEG-4 incorporates a syntax and various coding methods to support content-based manipulation and bit stream editing, without requiring transcoding (conversion from one coding system to another). This means that the user should be able to access a specific object within a scene or bit stream, thus making it possible to easily change the object's properties.

MPEG-4 offers efficient, content-based tools to access and organize multimedia data. These features include indexing, addition of hyperlinks, queries, viewing data, uploading or downloading data, and deleting data.

MPEG-4 supports efficient methods of combining synthetic and natural scenes (e.g., text and graphics overlays), the ability to code natural and synthetic audio and video data, as well as methods of mixing synthetic data with conventional video or audio under decoder control. This feature of MPEG-4 enables extensive interaction features. The hybrid coding of natural and synthetic data allows, for the first time, a smooth integration of natural and synthetic audiovisual objects and thus represents a first step towards the complete integration of all sorts of types of audiovisual information.

MPEG-4 supports scenes with different views/sound tracks and can efficiently code them together with sufficient information to synchronize the resulting basic streams. For video applications that use stereoscopic pictures or multiple views, this means that redundancy between multiple views of the same scene can be exploited. Furthermore, this permits solutions that are compatible with normal (mono) video. Coding multiple, simultaneous data streams should provide an efficient representation of natural 3-D objects if a sufficient number of views is available. On the other hand, this can necessitate a complex analysis process. This functionality should especially benefit applications that until now have used almost exclusively synthetic objects, for example in the area of virtual reality applications.

The strong growth of mobile networks in particular has resulted in a need for improved coding efficiency. MPEG-4 is thus needed in order to provide substantially better audiovisual quality than either existing standards or standards under development (e.g. H.263), at comparably low bit rates. It should be noted that the simultaneous support of other functionality does not necessarily further compression efficiency. However, this is not problematic since different coder configurations can be used in different situations. Subjective tests carried out with MPEG-4 in November 1995 showed that, in terms of coding efficiency, the available coding standards performed very well in comparison to most other proposed techniques.

Universal accessibility implies access to applications over a wide range of networks (both wireless and cable-based) and storage media. MPEG-4 must therefore work robustly in environments susceptible to errors. Especially for low bit rate applications, the system must provide sufficient robustness against serious errors. The approach used is not to replace the error-control techniques that are implemented in the network, but rather to offer elasticity against residual errors. This can be achieved with, for example, selective forward error correction, error containment, or error concealment.

MPEG-4 incorporates efficient methods to improve temporal random access to parts of an audiovisual sequence in a limited time interval with fine resolution. These include conventional techniques for achieving random access at very low bit rates.

7.7.5.2 Audiovisual Objects (AVOs) in MPEG-4

Audiovisual scenes in MPEG-4 are composed of audiovisual objects (AVOs), which are organized in a hierarchical fashion.

Primitive AVOs are found at the lowest level of the hierarchy. Examples are:

- a two-dimensional fixed background,
- the image of a talking person (without background) or
- the speech associated with the person.

MPEG-4 standardizes a number of these primitive AVOs that can be used to represent natural as well as synthetic types of content, which can be either two- or three-dimensional.

MPEG-4 defines the coded representation of such objects, for example:

- text and graphics,
- heads of speaking actors and associated text that the receiver can use to synthesize corresponding speech and animate the head accordingly,
- animated human bodies.

Figure 7-21 Example of an audiovisual scene in MPEG-4.

AVOs are individually coded in order to achieve maximum efficiency. They can be represented independently of other AVOs or of the background information.

MPEG is working with representatives of organizations that manage electronic media rights on defining a syntax for storing information about Intellectual Property Rights (IPR) pertaining to MPEG-4 AVOs and on developing tools to support IPR identification and IPR protection. The MPEG-4 standard thus incorporates support for unambiguous identification using attributes assigned by international naming authorities. These can be used to identify the current holder of the rights to an AVO. Protection of the content will be part of a subsequent version of MPEG-4.

7.7.5.3 Combining AVOs into Scenes

AVOs are combined in a hierarchical fashion to make audiovisual scenes. The result is a dynamic, tree-like structure of AVOs that can be changed through user interaction. Since each AVO possesses spatial and temporal components and can be located in relation to other AVOs, each AVO has coordinates that result from those of its parent AVO in the tree structure for the audiovisual scene.

7.7.5.4 Coding of Visual Objects

MPEG-4 uses a variety of methods to code visual objects. These are defined for natural images and videos as a set of tools and algorithms for the efficient compression of images, videos, textures, and 2-D and 3-D meshes, as well as geometry streams that animate these meshes in a time-varying manner. Tools to randomly access and manipulate all types of visual objects are available. Furthermore, content-based coding as well as content-based spatial, temporal, and qualitative scaling are supported. For natural content, there are mechanisms that provide robustness against errors and elasticity against errors in environments susceptible to errors.

To code synthetic objects, MPEG-4 defines parametric description tools as well as animated streams of human faces and bodies for static and dynamic mesh coding with texture mapping and texture coding. MPEG-4 uses wavelet compression techniques to efficiently code textures. Integration of other standards for synthetic audiovisual content, such as VRML, is planned as a future extension to MPEG-4.

In efficiently coding multimedia content the greatest gains can be gotten from video compression. Video coding is thus a particularly important aspect of the MPEG-4 standard. Three fundamental video coding extensions, which are realized by MPEG-4, are described below in detail:

1. Object-based scene layering and separate coding and decoding of layers

    In order to be able to support content-based functionalities, before processing a video scene, the coder must be able to divide the scene into layers that represent its physical objects. For example, one could divide a video scene into three layers O_1, O_2, and O_3, whereby O_1 represents the background, O_2 represents the person in the foreground, and O_3 represents the telephone (see Figure 7-22). Each object is then coded separately and transmitted over its respective bit stream layer.

Figure 7-22 Example of division of a video sequence.

The layering approach has the major advantage that each bit stream layer can be independently processed and separately coded and decoded. This achieves efficient coding and permits content-based functionalities.

If transmission takes place in environments susceptible to errors, different layers can be treated differently. For example, better error protection could be used for the foreground layer O_2 than for the two other layers if the recipient is more interested in the foreground object. Then at least the foreground layer can be decoded with sufficient quality if there is significant noise in the transmission channel. In other applications, the user could be interested in showing only the foreground layer when the video is being edited, manipulated, and mixed. The person in the foreground could then be placed in a scene coming from another video with a different background (possibly even a synthetic scene that was created through computer graphics). The bit stream layer O_3 can be accessed directly. This can also be inserted in the bit stream of another video scene without requiring that one of the two scenes be additionally segmented and transcoded.

2. Shape-adaptive DCT coding

After a scene has been successfully segmented, creating the different object layers, each layer is coded separately. This is done using a DCT coding technique that adapts itself to the object's shape. The fundamental structure of this technique can be considered an extension to conventional block-based hybrid DCT algorithms that use motion compensation and that code the different object layers. In contrast to previous standards, images to be coded in each object layer are no longer considered to be rectangular regions. Their shape and position can change between successive frames.

In order to model and optimize the performance of an algorithm for the functional coding of input sequences of arbitrarily shaped images at a bit rate of around 1 Mbit/s, the standard MPEG-1 coding system was extended with a shape coding algorithm and a shape-adaptive DCT technique. Shape-adaptive DCT allows transformation coding of image blocks of any shape.

For each object layer, the object shape information is transmitted first. This is followed by the blocks' motion vectors and the DCT coefficients corresponding to the motion and to the object's texture information as described by luminance and chrominance. In order to permit separate decoding of each object layer, the coding uses no information about shape, motion, or texture outside of the respective layer. As in previous MPEG definitions, input images to be encoded are split into grids of macro blocks and blocks. A block motion vector is transmitted for each macro block. For rectangular blocks, the SA-DCT algorithm reduces to the standard block DCT algorithm.

3. Object-based tool box for motion prediction

    The basic SA-DCT algorithm was developed in order to implement MPEG-4's object-based functionalities. The algorithm reduces temporal redundancy by using block-based motion compensation, comparable to that of the previous MPEG coding algorithm. An important advantage of the SA-DCT object-based approach is the significant increase in compression efficiency achieved by using appropriate tools for motion prediction in each object layer.

    The alternative prediction tools Pred 1–Pred 3, for example, extend the basic SA-DCT motion compensation approach with object-based motion prediction techniques. After the initial scene segmentation and layered description of the video content, each object layer can be classified in view of specific object properties during initial scene analysis (e.g., constant background with or without global camera movement motion, constant foreground objects with global motion or flexible foreground objects with related motion). This classification can be used to apply different tools for temporal prediction in layers with different properties. The set of motion prediction algorithms embedded in the basic SA-DCT coding system is also described as an object-based prediction tool box, in other words as a set of motion prediction tools that have been optimized for the motion statistics of various object layers having different properties in the object model.

    For flexible foreground objects with corresponding motion (e.g., the foreground person O_2) or for constant objects with global motion (e.g., a car) it might be more suitable to code and transmit fixed-motion vector fields or global motion parameters than block vectors. Moreover, a background with global camera motion can be coded very efficiently by estimating and transmitting global camera parameters, such as zoom, rotation, and translation, that can be mapped to a stationary panorama image of a constant background. This implies implementation within the coder and the decoder of a background memory that holds a complete (or as complete as possible) representation of the stationary background panorama. Based on these global parameters and on the panorama image stored in the background memory, a prediction of the background layer image to be coded is generated by the coder and by the decoder using efficient texturing algorithms.

Correction images for the background prediction are generated using the SA-DCT algorithm, similar to the hybrid DPCM/DCT standard coding method.

In the extended SA-DCT approach, conventional block-based motion compensation (Pred 1) can be used for object layers that do not satisfy the special model assumptions of the new prediction tools. In any case, intra-coded frames and regions where the model fails (for each of the prediction tools) are efficiently coded using the SA-DCT algorithm.

The MPEG-4 decoder was specified with different complexity levels in order to support different types of applications. These are: Type 0 (not programmable, with a predetermined set of tools), Type 1 (flexible, with a set of configurable tools) and Type 2 (programmable, with the ability to download new tools from the coder).

7.7.5.5 Streams in the MPEG-4 Standard

Just like MPEG-1 and MPEG-2, MPEG-4 also describes streams. Since MPEG-4 divides content into multiple objects, the stream properties affect multiplexing, demultiplexing, and synchronization of multiple streams.

AVO data is bundled into one or more Elementary Streams. These are characterized by the Quality of Service (QoS) needed for transmission (e.g., maximum bit rate or bit error rate) as well as other parameters, for example, stream type information, which helps determine the resources required in the decoder and the precision of time information in the coder. The manner in which such information about stream characteristics is transported from the source to the sink in a synchronized fashion over networks having different QoS is specified in terms of an Access Unit Layer and a conceptual two-layer multiplexer.

The Access Unit Layer allows Access Units to be identified within Elementary Streams, for example, video or audio frames, and scene description commands. This layer also permits the reestablishment of the time base of an AV-object or of a scene description as well as synchronization between them. The header of an Access Unit can be configured in a variety of ways, permitting a wide range of systems.

The FlexMux Layer (Flexible Multiplex) is fully specified by MPEG-4. It comprises a multiplexing tool that allows Elementary Streams (ESs) to be grouped with minimal multiplexing overhead. For example, Elementary Streams with similar QoS requirements could be grouped.

The TransMux Layer (Transport Multiplexing) models the layer that provides the transport services that match the required QoS. MPEG-4 specifies only the interface to this layer. Thus any suitable transport protocol architecture, such as (RTP)/UDP/IP/ (AAL5)/ATM or MPEG-2's Transport Stream over a suitable link layer, may become a specific TransMux instance. The choice is left to the end user/service provider, allowing MPEG-4 to be used in a multitude of operating environments.

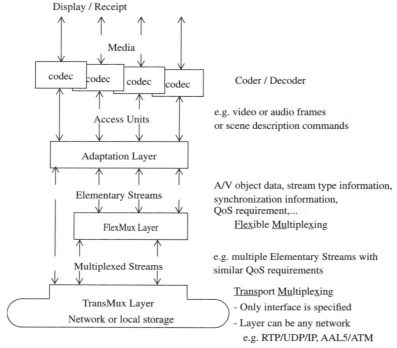

Figure 7-23 MPEG-4 System Layer Model.

Usage of the FlexMux multiplexing tool is optional. As shown in Figure 7-23, this layer can be skipped by the underlying TransMux instance since this provides equivalent functionality. The Access Unit Layer, however, is always present.

With this the following are possible:

1. Identification of Access Units, transport timestamps, clock reference information, and data loss.
2. Optional interleaving of data from different Elementary Streams into FlexMux streams.
3. Control information:
   - to indicate the required QoS for each Elementary Stream and FlexMux stream,
   - to translate such QoS requirements into currently available network resources,
   - to transport the mapping of Elementary Streams associated with AVOs to FlexMux and TransMux channels.

Individual Elementary Streams must be read from incoming data on a network connection or from a storage device. In the MPEG-4 system model, each network connection or file is homogeneously viewed as a TransMux Channel. Multiplexing is carried out partially or completely by layers that are not part of MPEG-4, depending on the application. The Stream Multiplex Interface is used as a reference point for integrating MPEG-4 in system environments. Streams packetized by the Adaptation Layer (AL) are accessed at this interface. The FlexMux layer specifies the optional FlexMux tool. The TransMux interface specifies how either streams packetized by the AL (no FlexMux used) or FlexMux streams are received from the TransMux layer. The TransMux interface thus realizes the passage to transport functionality not defined by MPEG. The interface's data part is considered here, while the control part is dealt with in the context of DMIF.

In the same way that MPEG-1 and MPEG-2 describe the behavior of an ideal operational decoding device together with the syntax and semantics of the bit stream, MPEG-4 defines a System Decoder Model. This allows the precise definition of a terminal's operation without having to make unnecessary assumptions about implementation details. This is essential in order to give developers the freedom to implement MPEG-4 terminals and decoding devices in a multitude of ways. Such devices range from TV receivers, which cannot communicate with the sender, to computers, which are fully capable of exchanging data bidirectionally. Some devices receive MPEG-4 streams over isochronous networks, while others use nonisochronous means (e.g., the Internet) to exchange MPEG-4 information. The System Decoder Model represents a common model that can serve as the basis for all MPEG-4 terminal implementations.

The MPEG-4 demultiplexing step is specified in terms of a conceptual two-layer multiplexer consisting of a TransMux layer and a FlexMux layer as well as the Access Unit Layer, which conveys synchronization information.

The generic term TransMux Layer is used to abstract underlying multiplexing functionality (existing or future) that is suitable for transport of MPEG-4 streams. It should be noted that this layer is not defined in the context of MPEG-4. Examples are the MPEG-2 Transport Stream, H.223, ATM AAL2, and IP/UDP. The TransMux Layer is modelled by means of a protection sublayer and a multiplexing sublayer, which make it clear that this layer is responsible for offering a specific QoS. The functionality of the protection sublayer includes error protection and error detection tools appropriate to the network or storage medium. In some TransMux instances, it may not be possible to separately identify the two sublayers.

Every concrete application scenario uses one or more specific TransMux Instances. Each TransMux Demultiplexer provides access to the TransMux Channels. Requirements for access to the TransMux Channel at the data interface are the same for all TransMux Instances. They include reliable error detection, the delivery or erroneous data with a suitable error indication (if possible), and the division of the payload into frames. This division consists of either streams that have been packetized by the AL or of FlexMux streams. The requirements are summarized in an informal fashion in the TransMux interface, in the system part of the MPEG-4 standard.

On the other hand, MPEG fully describes the FlexMux Layer. It provides a flexible tool for optional interleaving of data with minimal overhead and low delay and is particularly useful if the underlying TransMux Instance has a large packet size or high overhead. The FlexMux itself is not robust against errors. It can either be used on top of TransMux Channels with high QoS or to bundle Elementary Streams, both of which are equivalent in terms of error tolerance. The FlexMux requires reliable error detection and sufficient division of the FlexMux packets into frames (for random access and error correction), which must be provided by the underlying layer. These requirements are summarized in the Stream Multiplex Interface, which defines data access to individual transport channels. The FlexMux demultiplexer receives AL-packetized streams from FlexMux Streams.

The Access Unit Layer has a minimal set of tools for checking consistency, padding headers, conveying time base information, and transporting time-stamped access units of an Elementary Stream. Each packet consists of an access unit or fragment thereof. These time-stamped units represent the only semantic structure of Elementary Streams that is visible in this layer. The AU Layer requires reliable error detection and framing of individual packets of the underlying layer, which can for example be performed by the FlexMux. The manner in which the compression layer accesses data is summarized in the informal Elementary Stream interface, also in the system part of the MPEG-4 standard. The AU layer retrieves Elementary Streams from the streams packetized by the AL.

Depending on the degree of freedom allowed by the author of a scene, the user can interact with the scene's contents. The user could, for example, navigate through the scene, move objects to different positions, or trigger a sequence of events by clicking on a specific object with the mouse (for example, starting or stopping a video stream or choosing the desired language if multiple language channels are available). It would also incorporate more complex behaviors. For example, a virtual telephone rings and the user answers it, establishing a communications link.

Streams that come from the network (or from a storage device) as TransMux Streams are demultiplexed into FlexMux Streams and passed on to appropriate FlexMux demultiplexers, which receive Elementary Streams (see Figure 7-24).

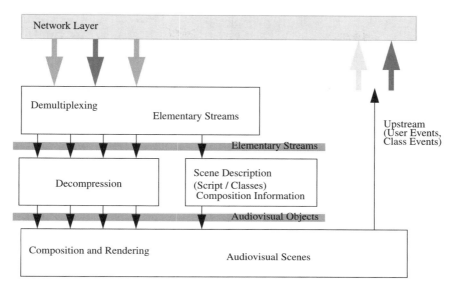

Figure 7-24 Important components of an MPEG-4 terminal.

Parts of the control functionality are only available in conjunction with a transport control unit, such as the DMIF environment, which MPEG-4 defines for precisely this purpose. The Delivery Multimedia Integration Framework (DMIF) covers operations of multimedia applications over interactive networks, in broadcast environments or from hard disks. The DMIF architecture is structured such that applications that use DMIF for communication do not need to have any knowledge of the underlying communications methods. The DMIF implementation takes care of the network details, presenting the application with a simple interface. DMIF can be placed between an MPEG-4 application and the transport network (see Figure 7-25).

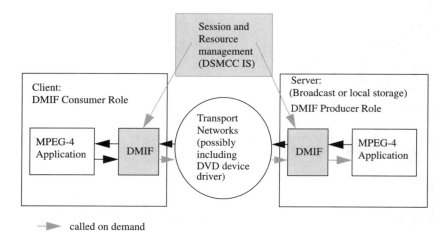

Figure 7-25 DMIF architecture.

To predict the behavior of a decoder during decompression of the various elementary data streams that make up an MPEG-4 session, the System Decoder Model makes it possible for the coder to specify and monitor the minimum buffer resources needed to decode the session. The required resources are conveyed to the decoder with object descriptors during establishment of the MPEG-4 session so that the decoder can decide if it is capable of proceeding with the session.

By managing the finite amount of buffer space, for example, the model allows the sender to transmit nonreal-time data too early, if sufficient space is available to store it at the receiver. The prestored data can be accessed at the appropriate moment, at which time real-time information can use a greater portion of the channel capacity, if required.

Real-time operation assumes a timing model where the end-to-end delay from the signal output of the coder to the signal input of the decoder is constant. Furthermore, the transmitted data streams must carry either implicit or explicit time information. There are types of such information. The first is used to convey the clock speed, or the encoder time base, to the decoder. The second, which consists of time stamps attached to parts of the coded AV data, contains desired decoding times for access units or the composition and expiration time for composition units. This information is conveyed in the AL-PDU headers generated by the Access Unit Layer. This timing information can be used to adapt the interval between pictures and the audio sampling rate at the decoder to match the encoder's values in order to achieve synchronized operation.

Different AV objects can be encoded by encoders with different time bases together with the slightly different speeds that occur. At any time, it is possible to map these time bases to the receiving terminal's time base. However, in this case no real implementation of a receiving terminal can avoid the occasional repetition or loss of AV data caused by temporal aliasing (relative reduction or extension of the time scale).

Although system operation without any timing information is allowed, it is not possible to define a buffer model in this case.

7.7.6 MPEG-7

After MPEG-3 did not materialize, it was decided not to allow MPEG-5 or MPEG-6 either, in order to preserve the logic of the sequence MPEG-1 (+1) to MPEG-2 (+2) to MPEG-4 (+3) to MPEG-7 (+4) ...

MPEG-7 is not meant as another compression format, rather the intention of the MPEG-7 ISO Working Group was to establish a metadata standard to supplement content coded in other formats (such as MPEG-4). The use of metadata aims at a new type of extension to the coding of multimedia data: mainly improved searching techniques and display strategies, but also consistency checking and, for example, scaling that incorporates consistency and priority. MPEG-7 calls for the integration and extension of other media content formats (especially MPEG-4) in this way.

In addition to the coding of metadata, MPEG-7 will define interfaces (and only the interfaces) for working in tandem with tools for automatic content analysis and search engines, but not these services themselves.

7.8 Fractal Compression

Fractal image and video compression represents an entirely different coding method. In this technique, no actual pixel information is transmitted, rather only a transformation function that contains an image similar to the target image as a fixed point. The decoding process then consists of an iterative application of this function to any beginning image.

Various specific attributes of this coding technique follow from this. First of all, the decoding process is progressive and the decoding efficiency is scalable, since the quality of the decoded image increases with each iteration step. Additionally, the process is independent of the resolution. The mapping function can be applied repeatedly in order to get more detail. The third attribute is an asymmetry between the encoding and decoding processes.

This raises the question of how an image can be coded as a transformation function. The algorithm makes use of a fractal attribute of images: their self-similarity. Images consist of regions that are similar to one another. A transformation function then consists of a mapping of each region of an image to the most similar part of the image. The mapping consists of contracting, stretching, rotating, and skewing the form of the image regions and adjusting their contrast and brightness. This is a type of vector quantization that does not use a fixed set of quantization vectors. The mapping will result in an image if the existence of a fixed point can be guaranteed. The only condition required by the theory for this is that of contractivity. The absolute value of the contraction factor must lie in the interval $[0, 1)$.

The mapping is achieved in current implementations [BH93] by dividing the original image into blocks (of 8×8 pixels) and, for each block, finding the most similar block of image regions (i.e., overlapping 16×16 pixel blocks). The set of blocks compared with each (8×8) original block is increased by the possible geometric transformations. The most similar block can be found by minimizing a distance measure. Usually the sum of the quadratic pixel differences is minimized.

For natural images, fractal compression can achieve high compression ratios (up to 1000:1) with very good image quality. The biggest disadvantage of this coding method is the complexity of calculation and its low efficacy when applied to graphics images. In order to keep the complexity within practicable limits, only a subset of all transformations are considered, for example only rotation angles of $0°$, $90°$, $180°$, and $270°$. Nevertheless each original block must be compared with a very large number of blocks in order to find the mapping to the most similar block. In addition to the complexity of calculation, this coding technique is lossy since it uses only the similarity of blocks, not their identity.

7.9 Conclusions

The important compression techniques used in multimedia systems all represent a combination of many well known algorithms:

7.9.0.1 JPEG

JPEG is the standard for still image coding that will have the most significance in the future. Its far-reaching definition allows a large number of degrees of freedom. For example, an image can have up to 255 components, that is, levels. An image can consist of up to 65,535 lines, each of which can contain up to 65,535 pixels. Compression performance is measured in bits per pixel. This is an average value calculated as the quotient of the total number of bits contained in the coded picture and the number of pixels contained in the picture. That said, the following statements can be made for DCT-coded still images [Wal91]:

- 0.25 to 0.50 bit/pixel: moderate to good quality; sufficient for some applications.
- 0.50 to 0.75 bit/pixel: good to very good quality; sufficient for many applications.
- 0.75 to 1.50 bit/pixel: excellent quality; suitable for most applications.
- 1.50 to 2.00 bit/pixel: often barely distinguishable from the original; sufficient for almost all applications, even those with the highest quality requirements.

In lossless mode, a compression ratio of 2:1 is achieved despite the remarkable simplicity of the technique. Today JPEG is commercially available in software as well as in hardware and is often used in multimedia applications that require high quality.

The primary goal of JPEG is compression of still images. However, in the form of Motion JPEG, JPEG can also be used for video compression in applications such as medical imaging.

7.9.0.2 H.261 and H.263

H.261 and H.263 are already established standards. These were mainly supported by associations of telephone and wide area network operators. Due to the very restricted resolution of the QCIF format and reduced frame rates, implementing H.261 and H.263 encoders and decoders does not cause any significant technical problems today. This is especially true if motion compensation and the optical low-pass filter are not components of the implementation, although the quality is not always satisfactory in this case. If the image is encoded in CIF format at 25 frames/s using motion compensation, the quality is acceptable. H.263 is mostly used for dialogue mode applications in network environments, for example video telephony and conferencing. The resulting continuous bit rate is eminently suitable for today's wide area networks operating with ISDN, leased lines, or even GSM connections.

7.9.0.3 MPEG

MPEG is the most promising standard for future audio and video compression use. Although the JPEG group has a system that can also be used for video, it is overly focussed on the animation of still images, instead of using the properties of motion pictures. The quality of MPEG video (without sound) at about 1.2 Mbit/s, appropriate for CD-ROM drives, is comparable to that of VHS recordings [Le 91]. The compression algorithm works very well at a resolution of about 360×240 pixels. Obviously, higher resolutions can also be decoded. However, at a resolution of, for example 625 lines, quality is sacrificed. The future of MPEG points towards MPEG-2, which defines a data stream compatible with MPEG-1, but provides data rates up to 100 Mbit/s. This significantly improves the currently available quality of MPEG-coded data.

MPEG also defines an audio stream with various sampling rates, ranging up to DAT quality, at 16 bit/sample. Another important part of the MPEG group's work is the definition of a data stream syntax.

Further, MPEG was optimized by making use of the retrieval model for application areas such as tutoring systems based on CD-ROMs and interactive TV. Embedding this optimization in MPEG-2 will allow TV and HDTV quality at the expense of a higher data rate. MPEG-4 will provide high compression ratios for video and associated audio and is furthermore an appropriate tool for creating whole classes of new multimedia applications. However, currently the complex and still very new MPEG-4 standard is little used in widespread commercial applications.

JPEG, H.263, MPEG, and other techniques should not be viewed as competing alternatives for data compression. Their goals are different and partly complementary. Most of the algorithms are very similar, but not identical. Technical quality, as well as market availability, will determine which of these techniques will be used in future multimedia systems. This will lead to cooperation and convergence of the techniques. For example, a future multimedia computer could generate still images using JPEG, use H.263 or MPEG-4 for video conferencing, and need MPEG-2 to retrieve stored multimedia information. This is, however, a purely hypothetical conception and is not in any way meant to prejudge future development or strategies for these systems.

CHAPTER 8

Optical Storage Media

Conventional magnetic data carriers in the form of hard disks or removable disks are traditionally used in computers as secondary storage media. These offer low average access time and provide enough capacity for general computer data at an acceptable price. However, audio and video data, even in compressed form, place heavy demands on available storage capacity. The storage cost for continuous media is thus substantial unless other data carriers are used.

Optical storage media offer higher storage density at a lower cost. The audio compact disc (CD) has been commercially successful in the consumer electronics industry as the successor to long-playing records (LPs) and is now a mass-produced product. Due to the large storage capacity of this technology, the computer industry has benefited from its development, especially when audio and video data are stored digitally in the computer. This technology is thus the main catalyst for the development of multimedia technology in computers. Other examples of external devices that can be used for multimedia systems include video recorders and DAT (Digital Audio Tape) recorders.

However, even the aforementioned mass storage media are not adequate to meet the data rates and especially storage requirements implied by current quality standards for multiple applications and data. This has led to the development of new media with many times the storage density of CDs or DAT tapes. The most important one, designated Digital Versatile Disc (DVD), is based on CD-ROM technology and was standardized by an alliance of multimedia manufacturers.

Actually integrating multimedia into systems that do not offer random access (e.g. magnetic mass storage such as DATs) is possible in some cases, though not easy [HS91, RSSS90, SHRS90]. This chapter thus focuses primarily on optical storage. Other data carriers are not considered here since most either do not have special properties with respect to integrated multimedia systems, or will be considered as components of a media server.

In the following, we explain the basics of optical storage media, followed by a brief discussion of analog and Write Once Read Many (WORM) systems. CD-ROM and CD-ROM/XA are explained based on CD-DA. Various other developments that pertain to multimedia include CD-I and the Photo CD, which are covered briefly. In addition to read only developments of the CD, the subsequently introduced CD-WO and CD-MO have been available for some time. This chapter concludes by comparing these CD technologies with one another and mentioning possible further developments.

8.1 History of Optical Storage

The video disc was first described in 1973 in the form of the Video Long Play (VLP). So far the video disc as a read-only medium has not been commercially successful although a large number of different write-once optical discs of different sizes and formats have been marketed. Most developments are based on analog technology, which can satisfy the highest quality standards at an acceptable cost.

About ten years later, at the end of 1982, the Compact Disc Digital Audio (CD-DA) was introduced. This optical disc digitally stores audio data in high-quality stereo. The CD-DA specification, drawn up by N.V. Philips and the Sony Corporation, was summarized in the so-called *Red Book* [Phi82]. All subsequent CD formats are based on this description. In the first five years after the introduction of the CD-DA, about 30 million CD-DA players and more than 450 million CD-DA discs were sold [BW90b].

The extension of the Compact Disc to storage of computer data was announced by N.V. Phillips and Sony Corporation in 1983 and introduced to the public for the first time in November 1985. This Compact Disc Read Only Memory (CD-ROM) is described in the *Yellow Book* [Phi85], which later led to the ECMA-119 standard [ECM88]. This standard specifies the physical format of a compact disc. The logical format is specified by the ISO 9660 standard, which is based on an industry proposal *(High Sierra Proposal)* and allows access using filenames and a directory.

In 1986, N.V. Philips and the Sony Corporation announced Compact Disc Interactive (CD-I). CD-I is described in the *Green Book* [Phi88], which includes, in addition to the CD technology, a complete system description. In 1987, Digital Video Interactive (DVI) was presented publicly. The primary emphasis in DVI is on algorithms for compression and decompression of audio and video data stored on a CD-ROM.

In 1988, the Compact Disc Read Only Memory Extended Architecture (CD-ROM/XA) was announced. N.V. Philips, Sony Corporation, and Microsoft produced a specification of digital optical data carriers for multiple media, which was published in 1989 at the CD-ROM conference in Anaheim, California [Phi89].

Since the beginning of 1990, we have seen the development of the Compact Disc Write Once (CD-WO) and the rewritable Compact Disc Magneto Optical (CD-MO), both specified in the *Orange Book* [Phi91]. Additionally, since the beginning of 1995 there has been the rewritable and erasable CD-RW (Compact Disc Read Write), also specified in the *Orange Book* [Phi91] (Part 3).

At the beginning of 1996, efforts to develop compact discs with higher storage densities began. This was initially carried out in a proprietary fashion by individual firms. At the end of 1996, the firms joined to form the DVD Consortium. The first DVD specifications became available by the middle of 1997, and the first devices by the end of 1997.

8.2 Basic Technology

In optical storage media, the underlying principle is that information is represented by using the intensity of laser light reflected during reading. A laser beam having a wave length of about 780 nm can be focused to a resolution of approximately 1 µm. In a polycarbonate substrate layer, there are depressions, called pits, corresponding to the data to be encoded. The areas between the pits are called lands. Figure 8-1 shows a sectional view through an optical disc, running lengthwise along a data track. The pits and lands are represented schematically in the middle of the figure.

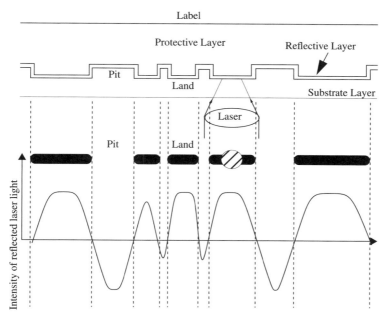

Figure 8-1 Sectional view of an optical disc along the data track. Schematic representation of the layers (top), the "lands" and the "pits" (middle), and the signal waveform (bottom).

The substrate layer is smooth and coated with a thin, reflective layer. The laser beam is focused at the height of the reflective layer from the substrate level. The reflected beam thus has a strong intensity at the lands. The pits have a depth of 0.12 μm from the substrate surface. Laser light hitting pits will be lightly scattered, that is, it will be reflected with weaker intensity. The signal waveform shown schematically at the bottom of Figure 8-1 represents the intensity of the reflected laser light; the horizontal line represents a threshold value. The laser in the figure is currently sampling a land.

According to Figure 8-1, a Compact Disc (CD) consists of:

- the label,
- the protective layer,
- the reflective layer and
- the substrate.

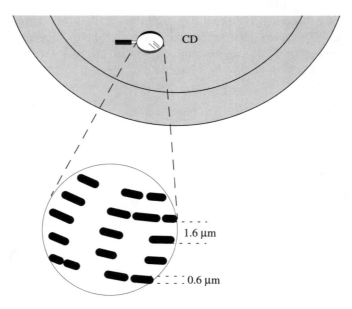

Figure 8-2 Data on a CD as an example of an optical disc. Track with "lands" and "pits."

An optical disc consists of a sequential arrangement of pits and lands within a track. The pits and lands represent data on the surface. Figure 8-2 shows a greatly enlarged detail of such a structure.

In contrast to floppy disks and other conventional secondary storage media, all the information on an optical disc is placed on one track. The stored information can thus be played back at a continuous data rate, which has particular advantages for continuous data streams such as audio and video data.

A track is in the form of a spiral. In the case of a CD, the spacing between adjacent coils of the spiral—the track pitch—is 1.6 µm. The track width of the pits is 0.6 µm, though their lengths can vary. The most important advantage of the CD over magnetic storage media follows from these measurements. Along the length of the data track, 1.66 data bits/µm can be stored, resulting in a storage density of 1,000,000 bit/mm^2. With the given geometry, this corresponds to 16,000 tracks/inch. In comparison, a floppy disk has only 96 tracks/inch.

While magnetization can decrease over time and, for example in the case of tapes, crosstalk can occur, optical storage media are not subject to such effects. These media are thus very well suited for long-term storage. Only a decomposition or change of the material can cause irreparable damage. However, according to current knowledge, such effects will not occur in the foreseeable future.

The light source of the laser can be positioned at a distance of approximately 1 mm from the disk surface and thus does not touch the disk directly, or float on an air cushion, as is the case with magnetic hard disks. This reduces wear and tear on the components used and increases the life of the device.

8.3 Video Discs and Other WORMs

Video discs in the form of LaserVision are used for the reproduction of motion picture and audio data. The data are stored on the disc in an analog-coded format, and the sound and picture quality are excellent. LaserVision discs have a diameter of approximately 30 cm and store approximately 2.6 Gbytes.

Following the naming of the well known long play (LP) for audio information, the video disc was originally called Video Long Play. It was first described in 1973 in the Philips Technical Review [Phi73].

Motion pictures are frequency-modulated on the video disc, and the audio signal is mixed with the modulated video signal. Figure 8-3 shows the principle used to record data. The important information of the mixed audio-video signal is the temporal sequence of the zero transitions. Each zero transition corresponds to a change between a pit and a land on the disc. Such a change can occur at any time, and is written to the disc in a nonquantized form, that is, the pit length is not quantized. This method is thus time-continuous and can be characterized as analog.

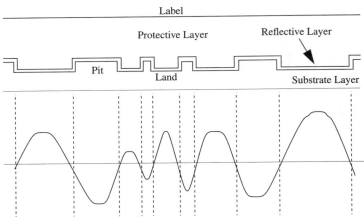

Figure 8-3 Sectional view of a video disc. Time-continuous discrete value coding.

The video disc was conceived as a Read Only Memory. Since then, many different write-once optical storage systems have come out, known as Write Once Read Many (WORM). An example is the Interactive Video Disc, which operates at Constant Angular Velocity (CAV). On each side, up to 36 minutes of audio and video data at 30 frames per second can be stored and played back. Alternatively, approximately 54,000 studio quality still images can be stored per side.

In 1992, there were already many write-once storage media with capacities between 600 Mbytes and about 8 Gbytes. These discs have diameters between 3.5 and 14 inches. The primary advantage of a WORM over rewritable mass storage is security against alteration. In order to further increase capacity, there are so-called jukeboxes, which can yield capacities of more than 20 Gbytes by using multiple discs.

Besides the large number of incompatible formats, software support is lacking in most systems. Computer integration is only available for a few selected systems.

WORMs have the following special properties:

- The term media overflow refers to problems that can occur when a WORM disc is almost full. Firstly, a check must be performed to see if data to be stored can fit on the WORM disc. Moreover, it must be determined if the data can or should be written on different physical discs, and if so, at what point in time to write to the new disc. This is particularly important for continuous media, since these data streams can only be interrupted at certain points.

- Packaging refers to problems stemming from the fixed block structure of WORMs. It is only possible to write records of one given size. For example, if the block size is 2,048 bytes and only one byte is written, then 2,047 bytes will be recorded without any information content.

- Revision refers to the problem of subsequently marking areas as invalid. For example, if a document is changed, the areas that are no longer valid must be marked as such, so that programs can always transparently access the current version. Such changes can result in a document being distributed over multiple WORM discs. Yet this must not disturb playback of continuous media streams.

8.4 Compact Disc Digital Audio

The Compact Disc Digital Audio (CD-DA) was developed jointly by N. V. Philips and the Sony Corporation for storing audio data. The basic technology of the CD-DA was developed by N. V. Philips [MGC82, DG82, HS82, HTV82].

8.4.1 Technical Basics

CDs have a diameter of 12cm and are played at a Constant Linear Velocity (CLV). The number of rotations per time unit thus depends on the radius of the data currently being sampled. The spiral-shaped CD track has approximately 20,000 windings; in comparison, an LP has only approximately 850 windings.

Information is stored according to the principle depicted in Figure 8-1 and Figure 8-4, whereby the length of the pits is always a multiple of 0.3 µm. A change from pit to land or from land to pit corresponds to the coding of a *1* in the data stream. If there is no change, a *0* is coded. In Figure 8-4, the data stream shown below corresponds to the coded data.

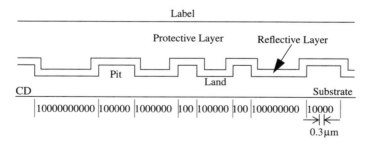

Figure 8-4 "Pits" and "lands." Discrete time, discrete value storage.

8.4.1.1 Audio Data Rate

The audio data rate can easily be derived from the given sampling rate of 44,100 Hz and the 16-bit linear quantization. The stereo audio signal is pulse code modulated, and the data rate is as follows:

Audio data rate$_{\text{CD-DA}}$

$$= 16 \frac{\text{bits}}{\text{sample value}} \times 2 \text{ channels} \times 44{,}100 \frac{\text{sample values}}{\text{s} \times \text{channel}} = 1{,}411{,}200 \frac{\text{bits}}{\text{s}}$$

$$= 1{,}411{,}200 \frac{\text{bits/s}}{8 \text{ bits/byte}} = 176.4 \frac{\text{Kbytes}}{\text{s}} \cong 172.3 \frac{\text{Kbytes}}{\text{s}}$$

Analog long-playing records and casette tapes have a signal-to-noise ratio of approximately 50 to 60 dB. The quality of the CD-DA is substantially higher. As a rule of thumb, one can assume 6 dB per bit used during the sampling process. Given 16-bit linear sampling:

$$\text{S/N}_{\text{CD-DA}} \cong 6 \frac{\text{dB}}{\text{bit}} \times 16 \text{ bits} = 96 \text{ dB}$$

The signal-to-noise ratio is exactly 98 dB.

8.4.1.2 Capacity

The play time of a CD-DA is at least 74 minutes. Using this value, the capacity of a CD-DA can easily be determined. The capacity given below applies only to storage used for audio data, without, for example, taking into account data used for error correction:

$$\text{Capacity}_{\text{CD-DA}} = 74 \text{ min} \times 1{,}411{,}200 \frac{\text{bits}}{\text{s}} = 6{,}265{,}728{,}000 \text{ bits}$$

$$= 6{,}265{,}728{,}000 \text{ bits} \times \frac{1}{8 \frac{\text{bits}}{\text{byte}}} \times \frac{1}{1{,}024 \frac{\text{bytes}}{\text{Kbyte}}} \times \frac{1}{1{,}024 \frac{\text{Kbytes}}{\text{Mbyte}}} \cong 747 \text{ Mbytes}$$

8.4.2 Eight-to-Fourteen Modulation

Each change from pit to land or land to pit corresponds to a channel bit of *1*; if no change takes place, the channel bit is a *0*.

Pits and lands may not follow each other too closely on a CD, since the resolution of the laser would not suffice to read direct pit-land-pit-land-pit... sequences (i.e., 11111 sequences) correctly. Therefore, it was agreed that at least two lands and two pits must always occur consecutively. Between every two *1*s as channel bits, there will thus be at least two *0*s.

On the other hand, pits and longs cannot be too long, otherwise a phase-correct synchronization signal (clock) cannot be derived. The maximum length of pits and lands was thus limited such that there can be at most ten consecutive *0*s as channel bits.

For these reasons, the bits written on a CD-DA in the form of pits and lands do not correspond directly to the actual information. Before writing, Eight-to-Fourteen Modulation is applied [HS82]. This transformation ensures that the requirements regarding minimum and maximum lengths are met.

Compact Disc Digital Audio

Eight-bit words are coded as 14-bit values. Given the minimum and maximum allowed distances, there are 267 valid values, of which 256 are used. The code table includes, for example, the entries shown in Table 8-1.

Audio Bits	Modulated Bits
00000000	01001000100000
00000001	10000100000000

Table 8-1 Eight-to-Fourteen code table. Two sample entries.

With direct sequencing of modulated bits (14-bit values), it is still possible that the minimum distance of two bits would not be met, or that the maximum distance of ten bits would be exceeded. Three additional bits are inserted between successive modulated symbols to ensure that the required regularity is met. These filler bits are chosen based on the neighboring modulated bits, as illustrated in Figure 8-5.

Audio Bits	00000000	00000001
Modulated Bits	01001000100000	10000100000000
Filler Bits	010	100
Channel Bits	01001001000100000100100000100000000	
On the CD-DA	1 ppp1 1 1 pppp1 1 1 1 1 1 ppp1 1 1 1 1 ppppppppp	

Figure 8-5 Integration of filler bits. *p* stands for *pit*, and *l* stands for *land*.

8.4.3 Error Handling

The goal of error handling on a CD-DA is the detection and correction of typical error patterns [HTV82]. An error is usually the result of scratches or dirt. These can usually be characterized as burst errors.

The first level of error handling implements two-stage error correction according to the Reed-Solomon algorithm. For every 24 audio bytes, there are two groups of correction data bytes, each of four bytes. The first group corrects single-byte errors while the second group corrects double-byte errors. The correction bytes also allow detection of additional errors in the sequence, although these cannot be corrected using this approach.

In the second level, real consecutive data bytes are distributed over multiple frames (a frame consists of 588 channel bits, corresponding to 24 audio bytes). The audio data are stored interleaved on the CD-DA. In this way, burst errors will always damage only parts of the data.

Using this technique, an error rate of 10^{-8} is achieved. Burst errors that span at most seven frames can be detected and corrected exactly. This corresponds to a track length of over 7.7 mm. In this case, no additional errors can occur in this area. One can, for example, drill a hole with a diameter of 2 mm in a CD-DA and still play the audio data correctly. However, experiments have shown that CD players do not correct all errors according to the given specifications.

The method described above for error correction is known as Cross Interleaved Reed-Solomon Code.

8.4.4 Frames, Tracks, Areas, and Blocks of a CD-DA

Frames consist of audio data, error correction, additional control and display bytes, and a synchronization pattern.

- The audio data are divided into two groups of 12 bytes each. They contain the high and low bytes of the left and right channels.
- Additionally, the error detection and correction bytes as described above are appended to each frame in two groups of four bytes each.
- Each frame has a control and display byte, consisting of eight bits, designated P, Q, R, S, T, U, V, and W (subchannels). Subchannel bits from 98 frames are collected and used together, yielding eight subchannels of 98 bits each, of which 72 bits are used for actual information. Every 98 frames form a block. Unfortunately, sometimes the blocks are also called frames. The *P* subchannel is used to distinguish a CD-DA with audio data from a CD with other computer data. The *Q* subchannel is used, for example:
    - in the lead-in area to store the directory,
    - in the rest of the CD-DA to specify the relative time within a track and the absolute time on the CD-DA.
- The synchronization pattern determines the beginning of each frame. It consists of 12 *1*s followed by 12 *0*s as channel bits and three filler bits.

Table 8-2 shows an overview of the components of a frame and its corresponding bits.

	Audio Bits	Modulated Bits	Filler Bits	Channel Bits
Synchronization			3+	24 =	27 bits
Control and Display		(14+3)		=	17 bits
12 Data	128	i.e.	12	(14+3)	=	204 bits
4 Error Handling		i.e.	4	(14+3)	=	68 bits
12 Data	128	i.e.	12	(14+3)	=	204 bits
4 Error Handling		i.e.	4	(14+3)	=	68 bits
Frame Total				=	588 bits

Table 8-2 Components of a frame.

Using these data, different data streams with corresponding data rates can be distinguished [MGC82]:

- The audio bit stream (also called the audio data stream) carries 1.4112×10^6 bit/s, counting only the 16-bit quantized sample values.
- The data bit stream includes the audio bit stream as well as the control and display bytes and bytes needed for error handling. The data rate is 1.94×10^6 bit/s.
- The channel bit stream contains the data bit stream with the Eight-to-Fourteen Modulation and filler and synchronization bits. The data rate is about 4.32×10^6 bit/s.

Altogether, a CD-DA consists of the following three areas:

- The lead-in area contains the directory of the CD-DA. The beginning of the individual tracks are registered here.
- The program area includes all tracks of the CD-DA; the actual data are stored here.
- At the end of each CD-DA, there is a lead-out area. This is used to help the CD player should it inadvertently read beyond the program area.

The program area of each CD-DA can consist of up to 99 tracks of different lengths. A CD-DA consists of at least one track, whereby a track usually consists of, for example, a song or a movement of a symphony. Random access to the beginning of each track is possible.

According to the *Red Book* specification, each track can have multiple index points, allowing direct positioning at certain points. Usually only two pre-defined index points (*IP*), IP_0 and IP_1, are used. IP_0 marks the beginning of each track, and IP_1 marks the beginning of the audio data within the track. The area within a track between IP_0

and IP_1 is called the track pregap. CD-DA discs have a track pregap of two to three seconds per piece.

Besides frames and tracks, another structure, called the *block,* was established (see Figure 8-6), though it does not have any significance for the CD-DA. In other CD technologies, it is used like a sector. A block contains 98 frames (see further details in Section 8.5.1).

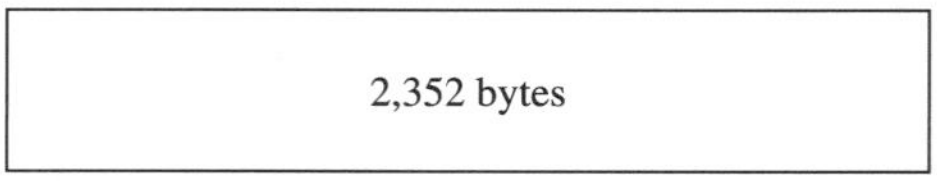

2,352 bytes

Figure 8-6 Actual data of a CD audio block (sector). Layout according to the *Red Book.*

8.4.5 Advantages of Digital CD-DA Technology

Errors on a CD-DA can be caused by damage or dirt. For uncompressed audio, the CD-DA is very insensitive to read errors that usually occur. As far as the digital technology, all CD-DAs are identical. An additional advantage is that there is no mechanical wear and tear.

The CD-DA specification as specified in the *Red Book* serves as the basis of all optical CD storage media. For example, Eight-to-Fourteen Modulation and the Cross Interleaved Reed-Solomon Code are always used. Hence, a fundamental specification was developed that is used in many systems, providing compatibility across the systems. However, the achievable error rate is too high for general computer data, necessitating an extension of the technology in the form of the CD-ROM.

8.5 Compact Disc Read Only Memory

The Compact Disc Read Only Memory (CD-ROM) was conceived as a storage medium for general computer data, in addition to uncompressed audio data [PS86, FE88, Hol88, LR86, OC89]. Further, CD-ROM technology was intended to form the basis for the storage of other media [KSN+87, Wil89]. It was specified by N. V. Philips and the Sony Corporation in the *Yellow Book* [Phi85] and later accepted as an ECMA standard [ECM88].

CD-ROM tracks are divided into audio (corresponding to CD-DA) and data types. Each track may contain exclusively data of one type. A CD-ROM can contain both types of tracks. In such a mixed mode disc (see Figure 8-13), the data tracks are usually located at the beginning of the CD-ROM, followed by the audio tracks.

8.5.1 Blocks

Since CD-ROMs store general computer data, they require better error correction and higher resolution random access to data units than are specified for CD-DA. A CD-DA has an error rate of 10^{-8} and allows random access to individual tracks and index points.

The CD-ROM data unit is called a block[1] and has similar properties to the sectors of other media and file systems. A CD-ROM block consists of the 2,352 bytes of a CD-DA block. Thus the *de facto* CD-DA standard can serve as the basis for the *de facto* CD-ROM standard.

Of the 2,352 bytes of a block, 2,048 bytes (computer data) or 2,336 bytes (audio data) are available for user data. The remaining bytes are used for identification for random access and for another error correction layer that further reduces the error rate.

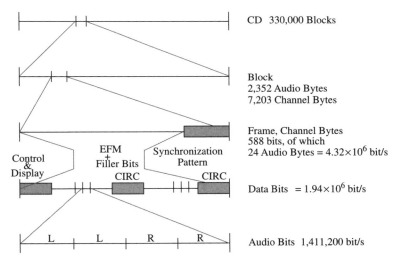

Figure 8-7 CD-ROM data hierarchy. Audio blocks as on a CD-DA.

Figure 8-7 shows the data hierarchy of a CD-ROM or CD-DA.

Seventy-five blocks per second are played back. Each block consists of 98 frames of 73.5 bytes (588 bits) each.

$$\text{Block} = 1{,}411{,}200 \frac{\text{bits}}{\text{s}} \times \frac{1}{75}\text{s} \times \frac{1}{8 \text{ bits/byte}} = 2{,}352 \text{ bytes}$$

---

1. This refers to the physical block; in ISO 9660, there is also the concept of a logical block.

8.5.2 Modes

The CD-ROM specification was defined with the goal of storing uncompressed CD-DA data and computer data, as well as serving as the basis for other media. This is achieved by using two CD-ROM modes. An additional mode 0, where all 2,336 user data bytes are set to zero, serves to separate storage areas.

8.5.2.1 CD-ROM Mode 1

CD-ROM mode 1 is used to store computer data, as shown in Figure 8-8. Of the 2,352 total bytes in each block, 2,048 bytes are available for storing information.

| Sync 12 | Header 4 | User Data 2,048 | EDC 4 | Blanks 8 | ECC 276 |

|←──────────────── 2,352 bytes ────────────────→|

Figure 8-8 CD-ROM mode 1 block (sector) layout according to the *Yellow Book*.

To be more exact, the 2,352 bytes can be broken down into the following groups:

- 12 bytes for synchronization as the start-of-block indicator,
- 4 bytes for the header. This contains an unambiguous block identifier. The first two bytes contain minutes and seconds, respectively; the third byte contains the block number, while the fourth byte identifies the mode,
- 2,048 bytes of user data,
- 4 bytes for error detection,
- 8 unused bytes, and
- 276 bytes for error correction, whereby an error rate of 10^{-12} can be achieved.

Given a playing time of 74 minutes, a CD-ROM can store 330,000 blocks.

The capacity of a CD-ROM with all blocks in mode 1 can be calculated as follows:

$$\text{Capacity}_{\text{CD-ROM}_{\text{Mode 1}}}$$

$$= 333{,}000 \text{ blocks} \times 2{,}048 \frac{\text{bytes}}{\text{block}} = 681{,}984{,}000 \text{ bytes}$$

$$= 681{,}984{,}000 \text{ bytes} \times \frac{1}{1{,}024 \frac{\text{bytes}}{\text{Kbyte}}} \times \frac{1}{1{,}024 \frac{\text{Kbytes}}{\text{Mbyte}}} \cong 650 \text{ Mbytes}$$

The data rate in mode 1 is:

$$\text{Rate}_{\text{CD-ROM}_{\text{Mode 1}}} = 2{,}048 \frac{\text{bytes}}{\text{block}} \times 75 \frac{\text{blocks}}{\text{s}} = 153.6 \frac{\text{Kbytes}}{\text{s}} \cong 150 \frac{\text{Kbytes}}{\text{s}}$$

8.5.2.2 CD-ROM Mode 2

CD-ROM mode 2 serves as the basis for additional specifications for storage of other media. A block in this mode is shown in Figure 8-9. Of the 2,352 total bytes in each block, 2,336 bytes are available for storing information.

Figure 8-9 CD-ROM mode 2 block (sector) layout according to the *Yellow Book*.

The synchronization and header are dealt with as in mode 1. The additional error correction is left out.

The capacity of a CD-ROM with all blocks in mode 2 can be calculated as follows:

$$\text{Capacity}_{\text{CD-ROM}_{\text{Mode 2}}} = 333{,}000 \text{ blocks} \times 2{,}336 \frac{\text{bytes}}{\text{block}} = 777{,}888{,}000 \text{ bytes} = 741.8518 \text{ Mbytes}$$

The data rate in mode 2 is:

$$\text{Rate}_{\text{CD-ROM}_{\text{Mode 2}}} = 2{,}336 \frac{\text{bytes}}{\text{block}} \times 75 \frac{\text{blocks}}{\text{s}} = 175.2 \frac{\text{Kbytes}}{\text{s}}$$

8.5.3 Logical File Format

It was recognized early on that the specification of blocks in mode 1 as an equivalent to other data carriers' sectors was not, by itself, sufficient to define a compatible data carrier, since the logical file format and the directory were missing.

A group of industry representatives thus met in Del Webb's High Sierra Hotel & Casino in Nevada and worked out a proposal that became known as the High Sierra Proposal. This proposal served as the basis for the ISO 9660 standard, which describes the format exactly (see, for example, its application in [KGTM90]).

The ISO 9660 standard defines a directory tree, which includes information about all files. In addition, there is a table that lists all the directories in a compressed form. This so-called path table allows direct access to files at any level. The table is loaded into the computer memory when a CD is mounted. Because a CD-ROM cannot be changed (read only), this method can be performed statically in an efficient fashion. However, most implementations use the actual directory tree.

ISO 9660 reserves the first 16 blocks (sectors 0 through 15) in the first track for the system area, which can be used in a vendor-specific manner. The volume descriptors start at sector 16 (for example, the primary volume descriptor or the supplementary volume descriptor). The most important descriptor is the primary volume descriptor, which includes, among other things, the length of the file system it defines as well as the length of the path table. Additional file systems can be defined using supplementary volume descriptors, which among other things offers flexibility with regard to the character sets allowed for file names.

Each volume descriptor is stored in a 2,048-byte block, and a CD-ROM can contain an arbitrary number of volume descriptions. Repeated copies of individual volume descriptors are usually stored in order to provide increased reliability in the case of a defective CD-ROM. The volume descriptor area ends with a volume descriptor terminator, which is implemented as a special block.

ISO 9660 established the logical block size as a power of two of at least 512 bytes. It may not exceed the size of the actual block (sector). The de facto maximum logical block size is 2,048 bytes, though this is not required by ISO 9660. If the underlying technology supports another physical block size, ISO 9660 allows other logical block sizes as well. Current block sizes are 512 bytes, 1,024 bytes, and 2,048 bytes. The logical block size is the same for the whole file system described by the volume descriptor. Files always start at the beginning of a logical block. Thus, files can begin and end within a block (sector). Directories, however, always begin on sector boundaries.

For some time there have been extensions to ISO 9660 explicitly supporting long file names, extended access rights, bootability, and special features of system-specific file systems. Among others, the following file system extensions have been introduced:

- Rockridge Extensions are used for specifying a version of the ISO 9660 format suitable for the Unix file system with long file names, links and access rights,
- The Joliet File System has been introduced by Microsoft, which implements extensions to adapt to the Windows 95/NT file systems and
- The El Torito Extension of the ISO 9660 system, which allows PC systems to boot directly from a CD-ROM.

8.5.4 Limitations of CD-ROM Technology

CDs have a high storage capacity and a constant data transfer rate. A random access time of about a second to an individual track can easily be tolerated for audio playback. This is a major improvement over CD audio or tapes. On the other hand, for a CD-ROM as a data carrier, these access times represent a significant disadvantage compared to magnetic disks (which have a mean access time of under 6 ms).

The following effects contribute to the time it takes to position to a desired block on a CD:

- Synchronization time occurs because the internal clock frequency must be adjusted to be exactly in phase with the CD signal. Here, delays are in the range of milliseconds.

- Due to the Constant Linear Velocity (CLV) playback of a CD, the rotational velocity at single speed is about 530 revolutions per second on the inside, but only about 200 revolutions per second on the outside. The rotation delay describes the time it takes to find the desired sector within a maximum of one rotation and to correctly set the rotation speed. Depending on the device, this time can be about 300 ms. For a CD-ROM device with a real 40-time data transfer rate and about 9,000 revolutions per minute, the maximum rotation delay is about 6.3 ms.
- The seek time refers to the adjustment to the exact radius, whereby the laser must first find the spiral track and adjust itself. The seek time frequently amounts to about 100 ms.

These partially overlapping effects cause a high maximum positioning time. However, real times can be highly variable, depending on the current position and the desired position. Using cache hierarchies, very good drives can reduce the access time to under 100 ms.

Outputting a steady audio data stream requires that audio blocks be stored sequentially on the CD. For example, an audio track cannot be played back at the same time data are read from a CD-ROM mode 1 track. Although this simultaneous retrieval is often very important for multimedia systems, it is not possible.

Today, there are already CD-ROM drives that sample the medium at speeds of up to 40 times that of a standard audio CD. This increases the data transfer rate achieved when reading large blocks. However, the access time during positioning, which is dominated by the seek time and does not depend directly on the rotational velocity, is not improved substantially.

8.6 CD-ROM Extended Architecture

The Compact Disc Read Only Memory Extended Architecture (CD-ROM/XA), which is based on the CD-ROM specification, was established by N.V. Philips, the Sony Corporation, and Microsoft [Fri92a, GC89, Phi89]. The main motivation was to address the inadequate consideration paid until then to concurrent output of multiple media. Prior to the CD-ROM/XA specification, this failing gave rise to other definitions and systems that included this capability. For example, there are the historically interesting CD-I (Compact Disc Interactive) and DVI (Digital Video Interactive) systems. Experience that N.V. Philips and the Sony Corporation obtained from developing the CD-I was incorporated into the development of the CD-ROM/XA. Many features of CD-ROM/XA and CD-I are thus identical.

The *Red Book* specifies a track for uncompressed audio data according to Figure 8-6. The *Yellow Book* specifies tracks for computer data using CD-ROM mode 1 (Figure 8-8) and tracks for compressed media using CD-ROM mode 2 (see Figure 8-9).

CD-ROM/XA uses CD-ROM mode 2 in order to define its own blocks and additionally defines a subheader that describes each block (sector) as shown in Figure 8-10 and Figure 8-11. This makes it possible to interleave different media using only mode 2 blocks, since these can contain different media. The individual CD-ROM/XA data streams are separated during playback.

Sync 12	Header 4	Sub-Header 8	User Data 2,048	EDC 4	ECC 276

|←──────────────── 2,352 bytes ────────────────→|

Figure 8-10 Sector layout (1) for CD-ROM/XA according to the *Green Book*. Data layout of a CD-ROM block in mode 2, form 1.

Sync 12	Header 4	Sub-Header 8	User Data 2,324	EDC 4

|←──────────────── 2,352 bytes ────────────────→|

Figure 8-11 Sector layout (2) for CD-ROM/XA according to the *Green Book*. Data layout of a CD-ROM block in mode 2, form 2.

8.6.1 Form 1 and Form 2

CD-ROM/XA differentiates blocks with form 1 and form 2 formats, similar to the CD-ROM modes:

1. The XA format form 1 in CD-ROM mode 2 provides improved error detection and correction. Like CD-ROM mode 1, four bytes are needed for error detection and 276 bytes for error correction. Unlike CD-ROM mode 1, the eight bytes unused in CD-ROM mode 1 are used for the subheader. Figure 8-10 shows a block (sector), where 2,048 bytes are used for data.
2. The XA format form 2 in CD-ROM mode 2 allows a 13 percent increase in actual data capacity, to 2,324 bytes per block, at the expense of error handling. Form 2 blocks can be used to store compressed data of various media, including audio and video data.

On a CD-DA, CD-ROM, or mixed mode disc, a track always consists of homogeneous data, meaning exclusively audio or computer data. It is thus not possible for the computer to, for example, concurrently read uncompressed audio data and computer data. The main advantage of CD-ROM/XA is that blocks of different media can be stored in one track since they are all coded in CD-ROM mode 2, thus allowing interleaved storage and retrieval.

8.6.2 Compressed Data of Different Media

Using interleaving, CD-ROM/XA allows the storage of different compressed media.

Audio can be compressed at different quality levels with ADPCM (Adaptive Differential Pulse Code Modulation). The total time of a CD-DA can be extended from 74 minutes (without compression) to over 19 hours of lower quality audio by reducing the audio signal to four bits per sample value. This compression is necessary in order to simultaneously retrieve data of other media. The following variants are possible:

- Level B stereo achieves a compression factor of 4:1 compared to a CD-DA audio signal. The level B sampling frequency is 37,800 Hz, resulting in a capacity of 4 hours and 48 minutes, based on a CD-DA playback time of 74 minutes. The data rate is about 43 Kbyte/s.
- Level B mono achieves a compression factor of 8:1 compared to a CD-DA audio signal. At this quality, 9 hours and 36 minutes of audio can be stored. The data rate is about 22 Kbyte/s.
- Level C stereo also achieves a compression factor of 8:1 and thus yields results in the same storage capacity and data rate as level B mono. The level C sampling frequency is 18,900 Hz.
- Level C mono works with a compression factor of 16:1, resulting in a maximum of 19 hours and 12 minutes with a data rate of about 11 Kbyte/s.

MPEG audio does not use ADPCM coding (see Section 7.7.2 regarding compression) and is thus still not compatible with the CD-ROM/XA specification. For other media, CD-ROM/XA is based on existing standards. Media-specific coding and decoding is not part of the CD technology. Thus, only references to other standards are given here.

When building applications using the CD-ROM/XA format, the maximum data rate must be considered when choosing the medium and the corresponding quality. The same applies to other CD-based formats, such as CD-I.

The logical format of CD-ROM/XA uses the ISO 9660 specification. ISO 9660 provides for interleaved files, that is, multiple files that are interleaved with each other. However, ISO 9660 does not address channel interleaving, that is, alternating sectors with audio, video, and other data within a file. ISO 9660 does not prescribe the content of a file. Unfortunately, the term file interleaving is often used for interleaved files as well as for channel interleaving.

An example of a system using this technology is a file system for PC systems under Unix described as early as 1992 [Kle92]. This file system is a component of the Unix kernel and is located under the Unix file system switch. All accesses that do not change data, even to other file systems, are possible this way. The implementation described uses CD-ROM/XA hardware with integrated chips for audio decompression.

In this system, low-resolution video can be decoded in software at 15 frames per second.

8.7 Further CD-ROM-Based Developments

The interaction of the different CD-ROM technologies is depicted in Figure 8-12. It should be stressed that the CD-DA, CD-ROM, and CD-ROM/XA specifications should be viewed like layers of a communications system.

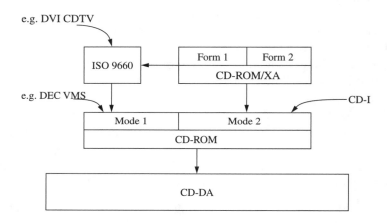

Figure 8-12 CD-ROM technologies. Specification of multiple layers.

Fundamentally, the CD-DA specification applies to all layers. However, the sequential development of these layers does not cover all the basic facts in the *Red Book*. For example, the mixed mode disc is not defined.

Other CD-based developments building on these fundamental technologies have appeared or are appearing to either handle multiple media or specific media and application areas. For the long term as well, it is expected that all further CD technologies will be based on CD-DA, CD-ROM, and CD-ROM/XA.

8.7.1 Compact Disc Interactive

The Compact Disc Interactive (CD-I) was developed by N. V. Philips and the Sony Corporation [vZ89] prior to the specification of CD-ROM/XA. In 1986, CD-I was announced. In 1988 the *Green Book* [Phi88] was defined based on the *Red Book* and the *Yellow Book* [Bas90, BW90a, B.V89, Sv91]. CD-I was originally designed for consumer electronics as an addition to the TV set, and the appropriate devices were available commercially in 1991. However, this system did not become widespread and disappeared entirely from the market by the end of 1997.

CD-I specifies an entire system. It includes a CD-ROM based format (different from CD-ROM/XA) with interleaving of various media and definition of compression for different media. Moreover CD-I defines system software based on the CD-RTOS operating system and output hardware for multimedia data.

The CD-I hardware is called the decoder. It consists of a main processor from the Motorola 68000 family together with special video and audio chips. It also includes a CD player with a controller and a joystick or mouse interface, and there is provision for a connection to a RGB monitor or a television. CD-I devices, which are the same size as video recorders, were intended to replace and extend CD-DA devices in the consumer environment.

The CD-I system software consists of the CD-RTOS operating system, a derivative of OS/9 with extensions for real-time processing.

8.7.1.1 Audio Coding

CD-I audio coding includes different quality levels with differing capacity and data rate. The different modes (CD-DA audio, A, B, and C) are listed in Table 8-3. The close relationship between CD-I and CD-ROM/XA is easy to see. CD-I was the basis for the CD-ROM/XA definition. The lower data rates can be used in combination with images or motion pictures. Multiple channels of lower quality can also be used for playback in different languages.

	CD-DA as comparison	CD-I Level A	CD-I Level B	CD-I Level C
Sampling Frequency in kHz	44.1	37.8	37.8	18.9
Bandwidth in kHz	20	17	17	8.5
Coding	16 bit PCM	8 bit ADPCM	4 bit ADPCM	4 bit ADPCM
Maximum Recording Duration in Hours (Stereo/Mono)	74 minutes/-	2.4/4.8	4.8/9.6	9.6/19.2
Maximum Number of Concurrent Channels (Stereo/Mono)	1/-	2/4	4/8	8/16
Portion (in %) of Total Data Stream (Stereo/Mono)	100/-	50/25	25/12.5	12.5/6.25
Signal-to-Noise Ratio (S/N) in dB	98	96	60	60
Equivalent Quality	Audio CD	LP	VHF Radio	Medium Wave Radio

Table 8-3 CD-I audio coding

8.7.1.2 Coding of Images

CD-I can be used to code images at different quality levels and resolutions. The following short overview shows that different data sizes and data rates are possible:

- The YUV mode is used for reproduction of natural images with many colors. The image resolution is 360×240 pixels, and the luminance component Y and the chrominance components U and V are coded using a total of 18 bits per pixel, allowing 262,144 colors per image. The resulting image size is:

$$\frac{\text{Data Size}}{\text{Image}} = 360 \times 240 \times 18 \times \frac{1 \text{ bits}}{8 \text{ bits/byte}} = 194{,}400 \text{ bytes}$$

- Using a Color Look-Up Table (CLUT), CD-I can work with four bits per pixel. Alternatively, three, seven, or eight bits per pixel are possible. This mode is suitable for simple graphics with fast retrieval using a preloaded color table. With four bits per pixel, at most 16 colors can be simultaneously presented. At a resolution of, for example, 720×240 pixels and four bits per pixel, the resulting image size is:

$$\frac{\text{Data Size}}{\text{Image}} = 720 \times 240 \times 4 \times \frac{1 \text{ bits}}{8 \text{ bits/byte}} = 86{,}400 \text{ bytes}$$

- The RGB mode is intended for very high-quality image output. Red, green, and blue components are each coded with five bits. Including an extra bit per pixel, colors are coded using a total of 16 bits per pixel, allowing up to 65,536 colors to per image to be displayed. With a resolution of 360×240 pixels, the data size for an image is:

$$\frac{\text{Data Size}}{\text{Image}} = 360 \times 240 \times 16 \times \frac{1 \text{ bits}}{8 \text{ bits/byte}} = 172{,}800 \text{ bytes}$$

8.7.1.3 Coding of Animations

Animations are run-length encoded using approximately 10,000 to 20,000 bytes per image. In the future, CD-I will use MPEG to code video. Although the CD-I file format was strongly influenced by the ISO 9660 standard, it is not completely compatible.

The CD-I technology was originally intended for the consumer marketplace. It is interesting in the context of the CD because it provided the basis for the CD-ROM/XA.

8.7.2 Compact Disc Interactive Ready Format

Although the different CD formats all stem from the CD-DA standard, it is not necessarily possible to play a CD-I disc on a CD-DA device. Furthermore, one cannot assume that all CD-DA devices will, for example, be replaced by CD-I devices. There was thus a need for a specification of an optical disc format that can be played on traditional CD-DA devices as well as on CD-I devices. This format is called the Compact Disc Interactive Ready Format [Fri92a].

In the Compact Disc Interactive Ready Format, the track pregap area between the index points IP_0 and IP_1 at the beginning of each track is increased from two to three seconds to a minimum of 182 seconds. The CD-I-specific information is stored in this area. This could be, for example, details about individual pieces, images, or biographies of the composer and the conductor.

A CD-I Ready Disc can be played in three different ways:

- With the usual CD-DA playback, the CD-I information in the track pregap is ignored and only the audio will be played.
- The second mode uses only the CD-I data in the track pregap. This can contain data of any media, which can be read, presented, and interpreted, possibly interactively. The CD-DA audio data of the track are not played.
- In the third mode, during the audio playback, the CD-I data from the track pregap are presented concurrently. This method is similar to the mixed mode disc (see Section 8.5). First, the CD-I data are read and stored. Then the audio information is output together with the corresponding data from the track pregap (which were read beforehand). In this way the data can be presented simultaneously.

8.7.3 Compact Disc Bridge Disc

The Compact Disc Bridge Disc (CD Bridge Disc)—like the CD-I Ready Disc—has the goal of enabling a CD to be output on devices supporting different formats. While CD-I Ready Disc specifies a disc format to allow output on either a CD-DA device or on a CD-I device, CD Bridge Disc specifies a disc format to allow output on either a CD-ROM/XA device or on a CD-I device [Fri92a].

Figure 8-13 shows the definitions of the aforementioned CD formats for output on devices using different standards.

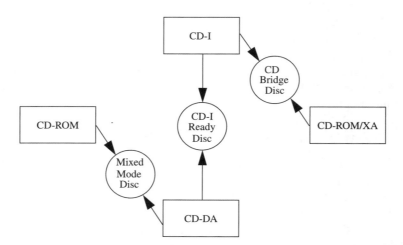

Figure 8-13 Compact Discs for output on devices with multiple formats. Mixed mode disc, CD-I Ready Disc, and CD Bridge Disc.

A CD Bridge Disc must satisfy both the CD-I and the CD-ROM/XA specifications, though it does not exploit all their capabilities. A common subset is defined that holds for both formats. All tracks with computer data (as opposed to uncompressed audio data as per CD-DA) must be recorded using CD-ROM mode 2. The disc may not contain any CD-ROM mode 1 blocks. Audio tracks (CD-DA) may follow the computer data tracks.

Another example of compatibility with both specifications is the track entry in the table of contents at the beginning of the CD. References to CD-I tracks are never included in this area. All tracks with data are thus marked as CD-ROM/XA tracks.

8.7.4 Photo Compact Disc

The Photo Compact Disc from Eastman Kodak and N.V. Philips is an example of a CD Bridge Disc [Fri92a] used for storing high-quality photographs. The Photo CD is based on CD-WO, so part of the Photo CD is delivered already written, and a second part can be written once. As a CD Bridge Disc, the Photo CD can be read by either CD-I or CD-ROM/XA devices. Additionally, it can be read and written by CD-WO devices and by special Photo CD devices.

The Photo CD was announced at Photokina '90 as the Kodak Photo CD System and will be licensed by Agfa-Gevaert.

The Photo CD is based on the following process [Kle92]: photographs are created using using conventional cameras and film. After the film is developed, the pictures are digitized using a resolution of eight bits for the luminance component and eight bits for each of two chrominance components. Each pixel is thus coded in 24 bits. Each photo is then coded in up to six resolutions as an ImagePac (see Table 8-4). This coding in multiple resolutions is analogous to the idea of hierarchical coding of images in JPEG (see Section 7.5.5 regarding compression). Each ImagePac usually requires 3–6 Mbytes of storage, depending on image complexity.

Image Version	Compressed/ Uncompressed	Number of Rows	Number of Columns
Base/16	uncompressed	128	192
Base/4	uncompressed	256	384
Base	uncompressed	512	768
4 Base	compressed	1,024	1,536
16 Base	compressed	2,048	3,072
64 Base	compressed	4,096	6,144

Table 8-4 Resolution of frames on a Photo CD.

The integration of photos with digital computer and television technology enables many new professional and consumer applications. For example, images can be displayed using a computer or TV. Using different resolutions, a digital zoom feature can be easily implemented. Multiple images can be displayed in an overview presentation by using a lower resolution. Using software, images can be modified after the fact and/or inserted into documents.

8.7.5 Digital Video Interactive and Commodore Dynamic Total Vision

Digital Video Interactive (DVI) specifies—like CD-I—different components of a system. DVI consists of compression and decompression algorithms; highly integrated, dedicated hardware components for compression and decompression of video in real time; a user interface (the Audiovisual Kernel, AVK); and a fixed data format. In contrast to CD-I, the emphasis is not on the CD technology, but on the compression algorithms [HKL$^+$91, Lut91, Rip89].

DVI uses CD-ROM mode 1 with the block format depicted in Figure 8-8. In addition, for the CD-ROM, DVI uses the ISO 9660 format as the basis for the AVSS (Audio/Video Support System) interleaved file format. Commodore's CDTV (Commodore Dynamic Total Vision), for example, also uses CD-ROM mode 1 and the ISO 9660 format. It should be noted that ISO 9660 distinguishes among different Interchange Levels. DVI uses the basic mode (Interchange Level 1), where among other restrictions, filenames are limited to 8-point-3 characters from a predefined character set. CDTV uses the Interchange Level 2, which allows, among other things, file names of up to 30 characters. Today neither system has any commercial significance.

8.8 Compact Disc Recordable

All of the CD technologies considered until now (except the Photo CD discussed in Section 8.7.4) do not allow users to write to discs, limiting the technologies' application domain. Research laboratories thus have developed and are developing, in parallel with read only storage media, compact discs that can be written once or multiple times.

The Compact Disc Recordable (CD-R), specified in the second part of the *Orange Book* [Phi91], like WORM (Write Once Read Many) allows the user to write a CD once and then read repeatedly [AFN90].

8.8.0.1 Principle of the CD-R

Figure 8-14 shows a sectional view through a CD-R vertically through the disc surface and the data track. The CD-R has a pre-engraved track. In all read-only CDs, the substrate (a polycarbonate) adjoins the reflective layer. In the CD-R, there is an absorption layer between the substrate and the reflective layer. Strong heat has the effect of irreversibly modifying this layer's reflective properties for laser beams.

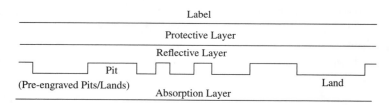

Figure 8-14 Sectional view of a CD-R disc, vertically through the data track. Schematic representation.

In the original state, a CD reader sees a track consisting of lands. Using a laser with an intensity of three to four times that of a read-only device's, the absorption layer in the area of the pre-engraved track is heated to above 250°C. This alters the material such that the reflected laser light now corresponds to a pit. This gives rise to the most noteworthy property of the CD-R. The data on a CD-R can be read by traditional devices designed exclusively for read-only CDs.

8.8.0.2 Sessions

All of the CD systems described so far assume that a lead-in area precedes the actual data area, and that this is followed by a lead-out area (see Section 8.4.4). The lead-in area holds a table of contents, which is needed by all playback devices in order to ensure correct positioning. However, when writing a CD-R this area cannot be written on the disc until after the write procedure has finished. Thus, all data on a CD-R would have to be copied to the disc in one atomic action. In the meantime, the disc would not be readable by other devices.

To solve this problem, the principle of multiple sessions, depicted in Figure 8-15, was introduced. Each session has its own lead-in and lead-out areas. In each write procedure, all the data for a session with its table of contents is written. The disc can then be read by other devices.

Figure 8-15 Layout of a "hybrid disc." Division into multiple sessions.

With this addition, the structure of a CD was extended to a maximum of 99 sessions. However, due to the space required for the lead-in and lead-out areas, at most 46 sessions—even with empty data areas—can be stored. In turn, each session consists of a lead-in area, a data area, and a lead-out area. Until 1992, all commercially available devices could read only one session. CD-R discs with only one session are called regular CD-R; those with more than one session are called hybrid CD-R.

CD-R recorders operate at up to eight times the data rate of a player. This shortens the write procedure, but also places requirements that cannot be ignored on the computer and additional software needed to produce a CD-R. This data rate must be sustained throughout the write procedure. Simpler programs thus first create an image of the CD-R on a hard disk. The data are then transferred to the CD-R in a second step. A storage-saving approach produces the data in the correct order and transfers them (without intermediate storage of the whole CD-R) at the necessary rate to the CD-R [Wep92].

Given the same price and features, the CD-R could, for example, be a substitute for CD-DA discs. However, the CD-R production process is and will continue to be more expensive than that of traditional CDs. Thus, it is used in other application areas. CD-Rs can be used whenever, for technical or legal reasons, large quantities of data need to be stored in an irreversible fashion. CD-R also finds application in the area of CD publishing because the expensive and time-consuming process of creating a master can be omitted. It is thus possible to produce limited-circulation editions that are very up to date.

8.9 Compact Disc Magneto-Optical

The Compact Disc Magneto Optical (CD-MO), specified in the first part of the *Orange Book* [Phi91], has a high storage capacity and allows the CD to be written multiple times.

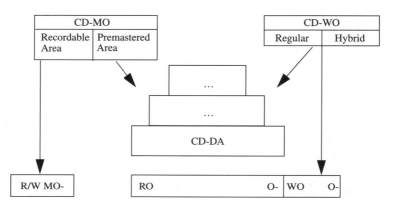

Figure 8-16 CD-WO and CD-MO in relation to other CD technologies: structure in multiple layers.

The magneto-optical technique is based on the principle that at higher temperatures, a weak magnetic field is needed to polarize the dipoles in certain materials. The block (sector) to be written is heated to above 150°C. At the same time, a magnetic field about ten times the strength of the Earth's magnetic field is applied. At this point, the material's dipoles are polarized towards this magnetic field. A pit is coded with a downwards-facing magnetic north pole. A land is coded using the opposite orientation.

In order to erase a block (sector), the area around the block is subjected to a constant magnetic field while it is heated.

If the CD is illuminated by a laser, the polarization of the light changes depending on the magnetization of the CD. In this way, the information can be read.

8.9.0.3 Areas of a CD-MO

A CD-MO consists of an optional read-only area and the actual rewritable area.

The read only area (premastered area in Figure 8-16) contains data written on the disc in a format specified for this purpose. In Figure 8-16, this relationship is indicated using the arrows between the premastered area of a CD-MO and the read only technologies. Thus, the read only area can be read using existing playback devices.

The rewritable area of a CD-MO cannot be read using existing CD-DA, CD-ROM, CD-ROM/XA, or CD-WO devices due to the fundamentally different technology used for reading and writing. Figure 8-16 shows the relationships between this recordable area and the underlying magneto-optical technology. This technology is thus incompatible with all the other CD technologies, although it was specified using the same system parameters as the other techniques. For example, the dimensions and the rotational speed are the same.

8.10 Compact Disc Read/Write

The Compact Disc Read/Write (CD-RW) is a further development of the CD-WO that, due to its physical makeup, can be repeatedly rewritten. This is achieved by using the reversible changeability of crystalline structures. The layers of a CD-RW are depicted in Figure 8-17.

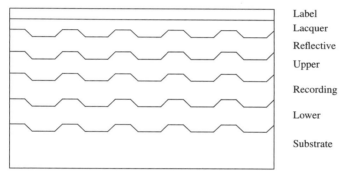

Figure 8-17 Layers of a CD-R/W. The upper/lower layers are protective layers that increase the stability of the CD-R/W.

Like the CD-R, phase changes are made by heating the crystal layer using the laser. However, in this case the energy is emitted as a pulse waveform (see Figure 8-18).

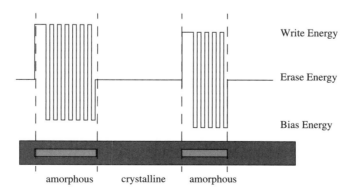

Figure 8-18 In the "pit" phases, the recording beam varies between the write energy and the bias energy in order to prevent overheating in the crystal.

However, using this technology it is no longer possible to read CD-RW discs on every CD player on the market since the reflectivity is lower than that of a CD or CD-R (see Table 8-5). In order to compensate for these effects, newer CD systems automatically amplify the signal as necessary. The other technology and logical structure are the same as the systems described earlier.

Type	Reflectivity
CD-DA	70%
CD-R/CD-WO	65%
CD-RW	15-20%

Table 8-5 Comparison of reflectivity of different CD variants.

8.11 Digital Versatile Disc

The *Digital Versatile Disc* (DVD) is, particularly in view of its larger storage space, the logical refinement of the CD-ROM/CD-R/CD-RW technologies. In 1994, large electronics companies began work on a CD with higher storage capacity. In early 1995, several companies joined to form the DVD Consortium and passed the first standards in the framework of the DVD Forum in April 1996.

8.11.1 DVD Standards

The DVD Consortium set down the specifications for DVD in the publications "Book A-E." Each Book defines a standard:

- DVD Read Only Specification (DVD-ROM, Book A): High capacity storage medium, successor to the CD-ROM,
- DVD Video Specification (DVD-Video, Book B): Specific application of DVD to distributing "linear" video data streams,
- DVD Audio Specification (DVD-Audio, Book C): Specific application of DVD to distribute pure audio data, similar to the CD-DA,
- DVD Recordable Specification (DVD-R, Book D): Variation of DVD that allows data to be recorded once for later use, and
- DVD Rewritable Specification (DVD-RW, Book E): Type of DVD that, like CD-RW, can be repeatedly written and erased; also known as DVD-RAM (Random Access Memory).

The storage capacity that can be achieved with DVD technology depends on the version (see Table 8-6).

Digital Versatile Disc

Version	Diameter (cm)	Sides	Layers per Side	Capacity (GB)	Comments
DVD-5	12	SS	SL	4.38	>2 hours video
DVD-9	12	SS	DL	7.95	about 4 hours video
DVD-10	12	DS	SL	8.75	about 4.5 hours video
DVD-18	12	DS	DL	15.9	>8 hours video
DVD-1*	8	SS	SL	1.36	about 1/2 hour video
DVD-2*	8	SS	DL	2.48	about 1.3 hours video
DVD-3*	8	DS	SL	2.72	about 1.4 hours video
DVD-4*	8	DS	DL	4.95	about 2.5 hours video
DVD-R	12	SS	SL	3.68	
DVD-R	12	DS	SL	7.38	
DVD-R	8	SS	SL	1.15	
DVD-R	8	DS	SL	2.3	
DVD-RAM	12	SS	SL	2.4	
DVD-RAM	12	DS	SL	4.8	

Table 8-6 DVD media variants and their storage capacities. SS: single sided; DS: double sided; SL: single layer; DL: double layer; * indicates term used here.

Here it is assumed that the standards listed in Table 8-7 are used for recording audio/video and data.

Video	ITU-T H.262/ISO-IEC 13818-2 (MPEG-2 VIDEO) ISO/IEC 11172-2 (MPEG-1 VIDEO)
Audio	ISO/IEC 13818-3 (MPEG-2 AUDIO) ISO/IEC 11172-3 (MPEG-1 AUDIO) Dolby AC-3-Standard
System	ITU-T H.222 / ISO/IEC 13818-1 (MPEG-2 System) program/ only PES Stream (no Transport Stream)

Table 8-7 DVD standards.

It should be noted that the capacity of a double-layer DVD is less than that of a double-sided DVD because the crosstalk that occurs when reading through the outer layer must be reduced.

8.11.1.1 Technical Basics

DVDs achieve a higher capacity than CD-ROMs by using smaller pits (which yields a higher track density), combined with a larger data area, more efficient coding of bits, more efficient error correction and lower sector overhead (see Figure 8-19).

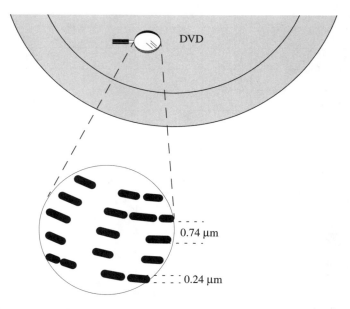

Figure 8-19 Data on a DVD. The track pitch and the pit width are less than a CD's.

From the standpoint of information technology, a DVD consists of a number of blocks of 37,856 bytes each. Each block contains 16 sectors plus additional data for error detection and correction. Individual sectors consist of 2,064 bytes divided into 12 rows as shown in Table 8-8. The first 12 bytes in the first row contain the sector header (sector ID, sector ID error correction, and six reserved bytes). The rest of the block, except the last four bytes, which contain the error detection code, holds user data.

Row	Contents
0	12 byte sector header and 160 bytes of user data
1	172 bytes of user data
...	
10	172 byte of user data
11	168 bytes of user data and four byte error detection code

Table 8-8 Structure of a DVD sector.

In order to transfer parallel streams better, DVD interleaves 16 sectors together. This also provides better robustness against errors. The result of the interleaving is a block of 192 rows (16 sectors × 12 rows per sector = 192 rows). At the end of each row ten bytes are added for further error correction, resulting in an additional 16 rows at the end of each block. Thus, only 33,024 bytes of each 37,856-byte block are available for user data, yielding a payload of only 87 percent.

8.11.2 DVD-Video: Decoder

In the following, some principles of DVD technology are illustrated by examining a decoder in the context of the DVD video specification. The decoder provides for the following six layers to transfer MPEG data:

- Layer 1: Synchronization, 8/16-demodulation, sector detection
  Altogether eight synchronization elements are inserted into the 8/16-coded bit streams. This layer recognizes and detects sector boundaries. At this step, the starting channel bit rate is 26.16 Mbit/s, while the ending user data rate amounts to 13 Mbit/s.
- Layer 2: Error detection (EDC) and correction (ECC)
  If the EDC check bits differ from the "fingerprint" generated from the data, then the inserted IEC data are used to help correct the error. After this layer, the user data rate is reduced to about 11.08 Mbit/s (about 2 Mbit/s are used for error correction, parity, and IEC data).
- Layer 3: Descrambling and decryption
  The data on the DVD are permuted in order to impede (or render impossible) unauthorized reading. Additional encryption is used for copy protection.
- Layer 4: EDC verification
  This is another error detection step.
- Layer 5: Track buffer
  The track buffer makes it possible to deliver the data that are read from the DVD at a fixed rate (11.08 Mbit/s) to an application at a variable data rate. Specific packets included in the data stream to control the player are dropped, yielding a maximum data stream of 10.08 Mbit/s.
- Layer 6: Transfer of data to MPEG System Decoder
  In this step, the data stream is demultiplexed into substreams that are delivered to their respective applications.

8.11.3 Eight-to-Fourteen+ Modulation (EFM+)

The lowest layer of the communication channel performs Eight-to-Fourteen+ Modulation (EFM+), U.S. Patent #5,206,646, which is mainly intended to reduce the DC component of the data stream. Like Eight-to-Fourteen Modulation, EFM+

eliminates certain sequences of bits (many zeroes). The main advantages of the 8/16 modulation are that no filler bits are necessary and that simpler decoding mechanisms are possible.

8.11.4 Logical File Format

The DVD file format (ISO 13490) is based on the ISO 9660 format. The ISO 13490 file format incorporates multisession capabilities specifically adapted to the features of the DVD technology.

8.11.5 DVD-CD Comparison

Table 8-9 provides an overview of the most important parameters of DVD technology compared with conventional CD technology.

	CD	DVD
Media Diameter	about 120 mm	120 mm
Media Thickness	about 1.2 mm	about 1.2 mm
Laser Wavelength	780 nm (infrared)	650 and 635 nm (red)
Track Pitch	1.6 µm	0.74 µm
Minimum Pit/Land Length	0.83 µm	0.4 µm
Data Layers	1	1 or 2
Sides	1	1 or 2
Capacity	about 650 MB	about 4.38 GB (SLSS) about 7.95 GB (DLSS) about 8.75 GB (SLDS) about 15.9 GB (DLDS)
Video Data Rate	about 1.5 Mbit/s	1-10 Mbit/s (var.)
Video Compression Standard	MPEG-1	MPEG-2
Video Capacity	about 1 hour	depending on format, between 2 and 8 hours
Sound Tracks	2-channel MPEG	2-channel PCM 5.1-channel AC-3 optional: up to 8 data streams
Subtitles	-	up to 32 languages

Table 8-9 Comparison of DVD with conventional CD technology.

8.12 Closing Observations

In the area of optical storage media, as far as can be seen today, Compact Disc technology or its further development in the form of the DVD, will predominate for all types of storage. The interrelationships among the different, mostly *de facto,* standards (see Figure 8-20) allow a wide field of application. With the exception of the CD-MO, the technologies are all based on the CD-DA and its optical technology.

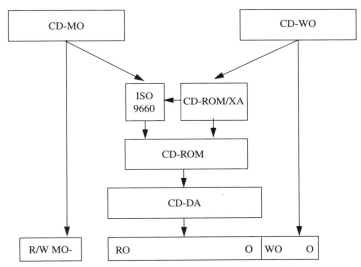

Figure 8-20 Important CD technologies and their relationships to one another.

A closer examination and comparison of the formats reveals the chronological sequence in which the CD technologies were specified. For example, CD-ROM in mode 1 defines improved error handling for computer data. CD-ROM/XA form 1, which is based on CD-ROM mode 2, provides the same service. It would thus be possible to eliminate CD-ROM mode 1 if there weren't already many applications that use it. The compression methods of CD-ROM/XA should permit use of the JPEG, MPEG, H.261, and CCITT ADPCM standards and not, for example, be limited to coding techniques supported by inexpensive chips already integrated on CD-ROM/XA controller boards.

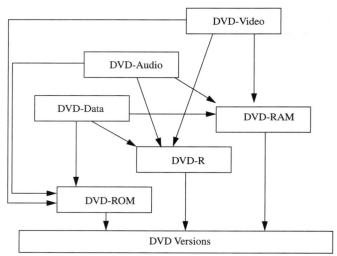

Figure 8-21 Important DVD technologies and their relationships to one another.

A major disadvantage of this technology is the relatively high average access time of at best about 200 ms. Even in the future, this will probably not be improved upon considerably. The data rate can be further increased by simultaneously reading multiple parts of the track that run parallel to one another. However, by then cache techniques may buffer the entire CD, so that the access time will no longer be determined by the CD device. The incompatibility between the CD and the CD-MO is unavoidable, but will become less important in the future due to CD-R/W and DVD-RAM technology.

The storage capacity achievable using CDs or DVDs is sufficient for many current systems. The storage density of up to 15.9 Gbytes obtainable today using DVDs is already sufficient for complex multimedia applications. With the quick progress in storage density, it is possible to imagine there will be further improvements in this area. As soon as stable robust semiconductor lasers with higher frequencies are available, a further jump in the storage density of optical media could follow.

The data rate of 10 Mbit/s obtainable with DVDs generally suffices to achieve very high quality audio and video playback with all currently used video and audio codings, such as MPEG-2 for video. However, this is not true for studio or movie theater quality. In this area, completely new developments are expected.

CHAPTER 9

Content Analysis

In recent years, multimedia documents have experienced wide distribution through the World Wide Web. In addition to the problems of receiving data over the network and storing it efficiently, users must now also cope with information overload. Search programs provide users looking for particular topics with so much information that they still must decide what is relevant. Information filtering tools, which provide the user with additional information about multimedia data besides just titles, are thus invaluable.

Most films currently available in digital form are uncommented and it is not to be expected that the large mass of existing films will ever be marked up "by hand" with a metadata track providing information about the actors or the film content. However, modern information technology is in principle capable of automatically extracting information from digital films, generating metadata that can be used to support users searching for specific content.

Pioneering works in this field include automatic cut detection in digital films [ADHC94, AHC93, ZKS93], automatic detection of newscasts [FLE95, ZGST94, ZS94], video indexing [GWJ92, RBE94, SC95, ZSW$^+$95, ZWLS95] and the extraction of key scenes from films [Ror93, LPE97]. These works are based on various areas of research in computer science [LWT94]: in addition to compression, pattern recognition, image recognition and signal processing make vital contributions to automatic content recognition of digital films.

This chapter begins with a discussion of currently available content analysis techniques inherent to various media. In this context, new types of applications have gained in importance; several of these are presented subsequently.

9.1 Simple vs. Complex Features

Features available for content analysis are called indicators and simple (syntactic) and complex (semantic) features are distinguished.

Syntactic indicators are features that can be extracted from a digital film through direct computation without any background knowledge regarding the content of the film. As a rule, it is not possible to directly make inferences about the content of the film by means of such indicators. Syntactic indicators are thus descriptive in nature. Only after they have been transformed into semantic indicators can a film be interpreted in such a way that, for example, its genre can be determined automatically.

Fundamentally two types of indicators can be defined:

- indicators that are valid at a fixed point in time t,
- and indicators that are defined over an interval of time.

For example, video indicators such as RGB color or gray value can be extracted from an individual image, while motion vectors can only be computed from a sequence of images.

Syntactic indicators thus represent both a transformation and an aggregation of a set of digital film material. Examples of syntactic video indicators include RGB color information, gray value information, information about color differences between images, edges in an image, similarity between images, motion vector information and segmentation of individual images into unicolored regions. Examples of syntactic audio indicators include volume, speech fundamental frequency, or the frequency distribution.

Semantic indicators allow film contents to be interpreted. Most of these features are obtained through a combined evaluation of previously considered syntactic indicators. In the video area, semantic indicators include zoom, cuts, fade-outs, dissolves, wipes, and the arrangement of segmented objects into logically related groups.

One differentiates between camera effects and editing effects. Camera effects are carried out as the film is shot, by the camera operator according to stage directions; editing effects are carried out when the film is edited. Camera effects include camera motion and zooms. A zoom is a camera operation whereby the camera operator enlarges or reduces the image in the direction of the zoom center.

In a dissolve between two scenes, the earlier scene is faded out while the following scene is simultaneously faded in. During this operation, the complete individual images of the superimposed scenes can always be identified.

In a wipe, the "fade to" (new) scene is superimposed over the currently visible scene such that there is no pixel overlap between images of the new scene and those of the previous scene. In the case of a horizontal wipe from left to right, left edge regions of images of the current scene are cut off and replaced with right edge regions of images of the new scene. The size of the region that is cut off is then successively enlarged until the new scene has completely replaced the old scene. Halfway through the time allotted for the wipe, the left half of the visible image is the right half of an image of the new scene, and the right half of the visible image is the left half of an image of the old scene.

In the audio domain, content-related features include, among many others, speech recognition; segmentation of an audio stream into discrete parts; and recognition of silence, music, and noises.

9.2 Analysis of Individual Images

Many features are available for the analysis of individual images. Such features include color, texture, objects in the image, and similarity between images or between edge images. These indicators can be used to perform classification tasks such as text recognition, which is described below. The analysis of individual images was already covered in detail in Chapter 4 (Graphics and Images).

9.2.1 Text Recognition

Examples of the need for processing information that is only available in written form on paper include the automatic recognition of addressee information in mail sorting systems, reading forms (e.g., bank transfer forms), transferring "old" data into an electronic form (e.g., converting a library's card catalog system into an online database), or handwritten text input into a PDA (Personal Digital Assistant). This section examines the automatization of these activities in greater detail.

In Optical Character Recognition (OCR) systems, text recognition takes place after an original document has been scanned in and is available as a bitmap [HHS96, SL95] (see Figure 9-1).

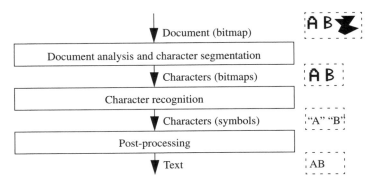

Figure 9-1 Text recognition in OCR systems.

9.2.1.1 Document Analysis and Character Segmentation

In the first step, the image is coarsely segmented into text and nontext regions. Regions that have a low probability of containing any text, for example, because they are made up of large unicolored areas, are discarded. In digital films, for example, subtitles can be recognized easily if one assumes that such text always appears in the lower fifth of the image and is oriented horizontally. Image enhancement is then performed on regions identified as containing text, for example, by removing underlining. In character segmentation, the text regions are divided into a series of individual characters. In the case of printed Roman characters, connected component analysis is frequently used since once assumes that, with few exceptions, individual characters consist of connected strokes. In the case of handwriting or Japanese characters, other methods must be used. These are not covered here.

9.2.1.2 Character Recognition

In character recognition, patterns representing individual characters are classified as characters of the underlying alphabet. A set of character features is determined and forms the input to the classification process. Two classification methods can be used: template matching and structural classification.

In template matching, each individual pixel of a character is taken as a feature. The pattern is compared with a series of stored character templates (possibly multiple templates per character, depending on the number of recognized fonts). A similarity measure is computed for each stored character template. Finally, the character is classified according to the template that yielded the highest similarity.

The structural classification method works differently. Here, individual structures within a character and their arrangement are analyzed, for example, for the letter "B," vertical or horizontal strokes, curves, or holes. The resulting structural features of a character to be classified are compared with known formation rules for each character of the alphabet.

9.2.1.3 Post-Processing

Connected words are then formed from the recognized letters and context information is used to correct recognition errors. For example, spell checking can be used to correct a sequence of letters recognized as "multirnedia" into the word "multimedia." In the case of recognizing destination addresses on letters, information about the postal code can be used to restrict the set of valid place and street names.

Current OCR systems achieve recognition rates above 99 percent when analyzing printed text.

9.2.2 Similarity-Based Searches in Image Databases

In many traditional image databases, images are described through manually collected text, for example, through a brief keyword content description or the name of the photographer. To query the database, these annotations are searched for specific user-provided keywords. This procedure allows for image description and searches, but is not always satisfactory since searches are limited to image features that appeared relevant to the annotator. Moreover, many image features are complicated if not impossible to put into words, for example, textures represented in an image. Furthermore, manual annotation is very time consuming.

It thus makes sense to develop image databases that support searches based on features that can be automatically extracted from the stored images, for example, color and texture or the presence of faces. It then becomes possible to make queries such as "Find images that have coloring similar to my example image" or "Find images in which three faces are present near one another." The output consists of a set of images from which the user must make a narrower selection. Currently available techniques are not advanced enough to allow general semantic requests, although expert systems can be used if the scope is limited to a specific type of images.

If a series of features has been computed for an image, these can be combined into a feature vector that describes the image. When the database is searched, the difference between each image's feature vector and the feature vector of the query image is determined by means of a distance function. The output set includes only images from the database that have a sufficiently small distance. If multiple features enter into the distance calculation, then the distance functions of the individual features must first be suitably normalized. For this one determines, based on results of psychological perception research, an interval $[a_1, a_2]$ over which human observers can discern distance values [LEJ98]. Distance values below a_1 are insignificant to human perception. Above a_2, saturation has been reached. Figure 9-2 shows an example of a normalization function.

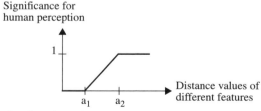

Figure 9-2 Normalization function.

For some features, such as color perception, this interval can be determined independently of a concrete query. For other features, individual determinations must be made, for example for the maximum permissible distance between human faces in the above query.

There are several ways to combine normalized feature distance functions into a total distance function. For example, one can use the so-called L_1 or L_2 metrics as a basis and introduce weights for the features. For two arbitrary feature vectors $m = (m_1, ..., m_n)$ and $m' = (m'_1, ..., m'_n)$, with weights $w_1, ..., w_n$ ($\Sigma w_i = 1$) and distance $d_i = |m_i - m'_i|$ the metrics can be computed as follows:

$$d_{L_1}(m, m') = \sum_{i=1}^{n} w_i d_i$$

$$d_{L_2}(m, m') = \sqrt{\sum_{i=1}^{n} w_i d_i^2}$$

If the covariance between the measures used is known, then the Mahalanobis distance

$$d_M(m, m') = \sqrt{\begin{bmatrix} m_1 & ... & m_n \end{bmatrix} C^{-1} \begin{bmatrix} m'_1 & ... & m'_n \end{bmatrix}^T}$$

can be used, which is recommended for strongly correlated measures (here C is the variance-covariance matrix) [Rus94].

9.3 Analysis of Image Sequences

Analyzing image sequences requires, besides analysis of individual images (also called *frames*), the analysis of their temporal structure. As a rule, the smallest logical data unit (LDU) above the frame level is the shot. A shot is a sequence of frames between two cuts that was recorded continuously by one camera. Multiple related shots form a scene, for example, a dialog in which two interview partners are shown alternately.

The following sections explain how to analyze the temporal structure of a film. Besides time, the feature of motion is essential. The explanation thus begins with techniques for computing motion vectors in films.

9.3.1 Motion Vectors

Motion is an important indicator for gleaning information from digital films. With the help of this indicator, for example, newscaster scenes can be separated from correspondents' news segments in a newscast. Unlike correspondent news segments, newscaster scenes exhibit a minimal amount of motion. In the following, several approaches to computing motion vectors are presented. These can be divided into the following categories:

- Block-oriented techniques determine objects' motion. Objects can either be fixed image blocks (e.g., 8×8 pixel blocks) or "real" objects extracted through segmentation.
- Pixel-oriented techniques take the movement of every individual pixel into account.

9.3.1.1 Block-Oriented Motion Vectors

In H.261 or H.263 and in MPEG-1 or MPEG-2 (see Chapter 7 regarding compression), block-oriented motion vectors have the great advantage that, in many practical applications, they do not need to be specially computed for content analysis purposes. In these MPEG encodings, which use P and B frames, the vectors are already available. Nevertheless, these vectors do not appear well suited as a basis for exact content analysis for the following reasons:

1. It cannot be assumed that a film to be analyzed was compressed using MPEG or H.261. Other compression formats such as wavelets generally do not compute any motion vectors.
2. In MPEG and H.261, actual motion within blocks cannot be recognized using the block-oriented motion vectors.

The last point alone is a serious disadvantage. Consider an image with a square moving towards the right (see Figure 9-3). Motion vectors computed by MPEG-1 or MPEG-2 and by H.261 or H.263 are only nonzero for edge regions where the difference between both images is calculated as nonzero. This is because the color value does not change between the two images for other areas. Since only block-oriented vectors (and no object-oriented vectors) are used, it is not possible to derive actual camera or object movement. It is thus only reasonable to use this motion analysis technique as a first approximation.

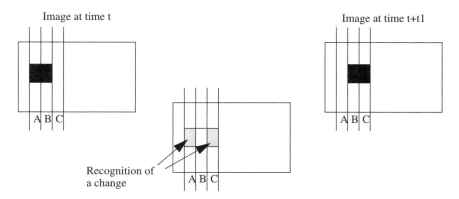

Figure 9-3 Block-based motion vectors.

9.3.1.2 Pixel-Oriented Motion Vectors

Techniques for computing motion vectors on a pixel basis are called optical flow methods.

Object movements in the real world appear in an image as color changes. In some circumstances, color changes in two successive images can indicate movements (see [Jäh97]).

As already described in Chapter 4 (Graphics and Images), tristimulus color values can be transformed into gray values. The advantage of gray values is that they can be differentiated in one-dimensional space, whereas color values span a three-dimensional space. Thus operations are generally much simpler to perform using gray values. In the following it is thus assumed that images for which motion vectors are to be computed are available as gray value images. The same technique can be used with color values, although it should be considered that in practice the computational time is significantly higher.

Optical flow refers to the movement of gray value patterns about the image area. First, the displacement vector is determined at each point for the respective gray value. Then a continuous vector field is computed, which should adequately reproduce the optical flow. Implementing both steps requires certain limiting assumptions and the results cannot be completely error-free [Jäh97].

Nevertheless, important spatio-temporal information can be obtained. It does not matter whether changes are continuous, caused by a change in vantage point, or discontinuous, due to individual objects. In any case, it should be clear that it does not make sense to consider individual images in isolation. A sequence of at least two successive images must be examined [Jäh97].

The following are classes of methods for computing optical flow [BFB94]:

- Differential methods, which use derivatives of images' gray values to determine motion vectors [Luc84, LK81, Nag83, Nag89, UGVT88].
- Correlation-based methods, which determine vectors using correlation between regions of successive images [Ana87, Ana89, Sin90, Sin92].
- Energy-based methods, which use the energy resulting from the application of velocity-dependent filters [Hee87, Hee88].
- Phase-based methods, which define velocity as a phase dependence of the response of bandpass filters [FJ90, WWB88].

When three-dimensional reality is projected onto a two-dimensional image plane, it is no longer possible to easily distinguish objects and background from one another due to the loss of the third dimension, particularly if there is no clear difference in the gray values. As described below, this has several interesting consequences for motion determination, which is based on recognizing and computing gray value differences in successive images in a sequence.

We start by assuming that gray value differences have been detected in an image pair. These are not necessarily the result of object movements. For example, the differences could have been caused by changing lighting conditions, by camera noise, as well as by a change in the camera position. In the latter case, the entire image would move, which would likely lead to difficulties at edges of the image, which are no longer identical. Here gray value changes occur without object movement. There is a physical correspondence between the images, although it is no longer understandable visually [Jäh97].

On the other hand, even if one considers objects with a uniform surface structure and coloring that rotate about an axis perpendicular to the image plane, under some circumstances gray value differences can occur that do not suffice to infer object movement. In this case, there is movement without (sufficient) gray value change [HS81].

Images always capture only a section of the real world, and the images themselves are viewed as through an aperture. If the aperture is shifted, or if an observed object moves, it is possible that there will be difficulties in correctly analyzing the image content. Gray value changes along an edge that extends across the entire image section can only be discerned in directions perpendicular to the edge. Movements which are parallel to the edge direction are present in the real world, but do not manifest themselves in gray value changes. Examples of this are shown in Figure 9-4. In each case, lines are recognized that were visible at times t_0 and t_1 in the respective images.

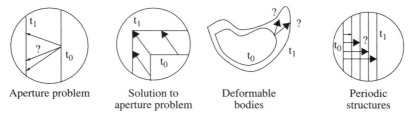

Figure 9-4 Optical flow problems.

In some cases, a nondeterminable motion component in the edge direction must be added to the computed displacement vector. This ambiguity is known as the aperture problem. Figure 9-4 illustrates this situation.

The aperture problem can obviously be solved if the image section is moved far enough that both images being examined contain either the end of the gray value edge or a corner of any angle [Jäh97]. It is then possible to unambiguously match the edges between the two images.

Additional problems occur when deformable bodies are considered. Here, an original point cannot be unambiguously mapped to its displaced point.

Even rigid bodies can lead to ambiguous computation results if they have periodic patterns or structures. The displacement vector can then only be determined to an accuracy of the order of a multiple of the width (length) of the repeated pattern or structure.

A third problem example is indistinguishable objects that move independently of one another. If these occur in large numbers, noncorresponding pairs cannot be uniquely identified. Visual correspondence (indistinguishability) provides the illusion of physical correspondence [Jäh97] (Figure 9-5).

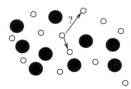

Figure 9-5 Physical correspondence.

With additional assumptions it is possible to find implementation approaches that resolve the indistinguishability problem. For example, one can assume that images are recorded at intervals short enough that movement by more than the structure or pattern width is not possible.

By assuming that indistinguishable particles move at approximately the same speed, it is possible to compute a solution with the smallest error by determining the variance of all possible associated pairs [Jäh97].

In a theoretical and experimental comparison, Barron, Fleet, Beauchemin, and Burkitt [BFB94] show that differential methods, in particular those of Lucas and Kanade [LK81] and Horn and Schunck [HS81] yielded the best results for processing digital film sequences at the time of the study (1994). Even today these methods yield outstanding results. The authors further show that of the correlation methods, the method ascribed to Singh [Sin90] yields the most reliable results.

9.3.2 Cut Detection

An important technique for dividing digital films into semantic units is the detection of editing effects, such as cuts, fade-ins and fade-outs, or dissolves and wipes [AJ94]. A cut denotes the sudden change of the image content from one image to the next. A scene consists of content-related shots, film segments demarcated by editing effects (see also [BS83, DB81, HJW95, Kuc87]).

Commercial applications of automatic digital cut detection are particularly prevalent in the area of video indexing systems [HJW94a, TATS94, ZKS93].

Well known methods for automatically recognizing editing effects are:

- methods that recognize editing effects based on histogram changes [NT91, OT93, RSK92, Ton91, ZKS93],
- methods that recognize editing effects with the aid of edge extraction [MMZ95],
- methods that recognize editing effects by means of chromatic scaling [HJW95], and
- methods that infer cuts from changes in the distribution of DCT coefficients [AHC93].

In the following, the term cut detection is used synonymously for the detection of cuts and other editing effects.

9.3.2.1 Pixel-Based Cut Detection

Pixel-based methods compute a difference between two images based on the images' pixel differences (the pixels can be either separate color values or gray values). The computation can be performed either through a pair-wise comparison of pixels or image blocks or through histogram comparison.

In the pair-wise comparison method, the pair-wise difference between two successive images is calculated. Differences occur if the color value change or gray value change of a pixel exceeds a threshold value T. In a pair-wise comparison, the number of pixels that change from one image to the next are counted. A cut is detected if a specific change percentage Z is surpassed.

The problem with this approach is that it is very sensitive to camera motion or object movement, causing these sorts of image change to be falsely recognized as cuts.

9.3.2.2 Likelihood Ratio

Here, instead of comparing individual pixels, image regions in consecutive images are compared using second order statistical methods [KJ91]. Cuts are automatically detected if an image has too many regions whose likelihood ratio exceeds a threshold value T. If m_i and m_{i+1} denote the mean intensity of a given region in two images and S_i and S_{i+1} denote the corresponding variances, then the likelihood ratio is defined as follows [KJ91]:

$$L = \frac{\left(\frac{S_i + S_{i+1}}{2} + \frac{(m_i - m_{i+1})^2}{2}\right)}{S_i \times S_{i+1}}$$

Compared to the method of pair-wise comparison, the likelihood ratio method has the advantage that small object movement or camera motion does not distort the results of the computation. Consequently, the total variance of the function is lower, which makes it easier to choose the threshold value. A potential problem is that two regions that have different distribution functions will be considered identical if they have the same mean value and the same variance. This case, however, is very rare.

9.3.2.3 Histogram Comparisons

Instead of comparing individual pixels or regions, the statistical distributions of two images' pixels can be compared. This is achieved by generating histograms that capture the distribution of gray values of an image. The idea here is that images in which objects and the background change only slightly will have similar gray value histograms. This technique can also be applied to RGB color images by using three-dimensional color histograms. However, this requires considerable computational expense, and the results are not improved substantially. Histogram-based algorithms are not sensitive to camera motion or object movement.

A potential problem is that two completely different images can have identical histograms. Figure 9-6 shows four different images that have the same histogram. The images' common histogram is shown in Figure 9-7. However, in practice this occurs so rarely between two adjacent frames in a film that this case need not be viewed as problematic.

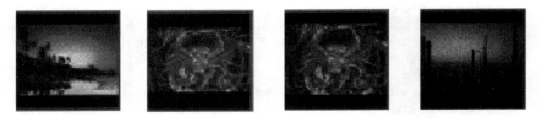

Figure 9-6 Images with identical histograms.

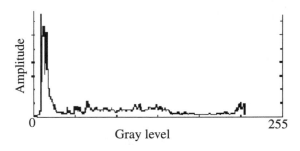

Figure 9-7 Histogram of images in Figure 9-6.

A cut is recognized if the histogram difference exceeds a specific threshold value T.

Figure 9-8 and Figure 9-9 show three consecutive images and their histograms. While the first two images have almost identical histograms, the histogram of the third image is markedly different. A cut is thus detected here.

Figure 9-8 Original images with cut after the second image.

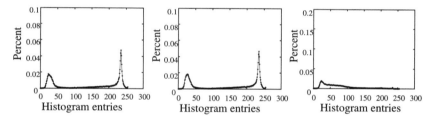

Figure 9-9 Histograms of images from Figure 9-8.

A problem with all approaches is that sudden lighting changes or deformable objects, which result in strong histogram changes, lead to false detections. Flashing lights or explosions, for example, are frequent error sources.

Nagasaka and Tanaka propose a technique to avoid these errors [NT91]. They observe that at most half the image changes, if lighting changes are excluded. The image can thus be divided into 16 rectangular regions, which are examined separately. The cut detection is then computed using the eight lowest comparison values.

9.3.2.4 Detection of Fade-Ins/Fade-Outs and Dissolves

Figure 9-10 shows the histogram differences of frames in a film scene that contains a cut and a fade-out. Note that the difference measure increases markedly for the cut, which is easily recognized by choosing an appropriate threshold value. On the other hand, the fade-ins and fade-outs show relatively minor increases in the difference measure. They cannot be recognized by simply reducing the threshold value, since many transitions would be falsely identified as cuts. In this context, transitions that are detected although none is present in the film material are called false positives.

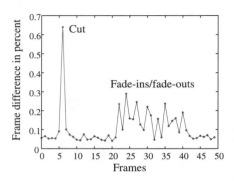

Figure 9-10 Transition detection using pixel image differences.

Observing that the first and last images of a transition are usually very different allows the problem to be reformulated as follows. Instead of searching for the transition effect, one can search for the beginning and ending of the transition. In order to accomplish this, two threshold values T_1 and T_2 must be used. As before, T_1 specifies the threshold where a cut is recognized. T_2 specifies a threshold value that marks the beginning of a potential transition. Then the difference between the marked image and the current image is computed until either T_1 is exceeded, meaning a gradual transition was found, or the difference again drops below T_2, at which point the marking of the image is ended. By selecting an appropriate tolerance value, it is possible to determine how often the difference value may fall below T_2 without the first image being unmarked [ZKS93].

A problem with this method is that the authors do not provide a way of differentiating a fade-in or fade-out from a dissolve or a wipe.

9.3.2.5 Cut Detection Based on Edge Extraction

Unlike histogram-based methods for cut detection, methods based on edge extraction make use of local information. Here the number of edge pixels that disappear or appear from one image to the next is measured (for edge extraction see Chapter 4). The individual steps of this method are as follows:

1. Calculation of the binary edge images E_i and E_{i+1} from two consecutive images I_i and I_{i+1}. In the edge images, a one represents an edge point and a zero represents any other image point.
2. Elimination of camera motion through motion compensation.
3. Calculation of the number of edge points that disappear or appear from one image to the next.
4. Decision as to whether or not a transition as occurred, whereby the measure used is the maximum of the number of edge pixels that disappeared or appeared.

Correlation-based methods are used for motion compensation between the images I_i and I_{i+1}; these include the Hausdorff distance [HKR93] and the census transform [ZW94]. Using these methods it is possible to calculate the camera motion. The Hausdorff distance is a displacement vector from which it is possible to derive the direction and length of the camera motion. If the image is simply moved by this vector, one finds that at most points, the images are not congruent, even excluding points that have newly appeared in the image due to object movement or camera motion. This process is also called warping [Wol90].

It is easy to detect transitions by looking for local maxima of the function that measures edge pixels that have newly appeared or disappeared from an image. If this exceeds a threshold value T, a transition is recognized. One must then decide whether the transition is a cut, a fade-in or fade-out, a dissolve, or a wipe. This depends on the threshold value used and on the time horizon over which the transition takes place.

A cut is easy to recognize, since in this case the time horizon is one image. Said another way, a cut causes a single deflection in the edge change function, whereas other transitions take place over greater intervals.

Fade-ins or fade-outs and dissolves can be differentiated from one another by considering the ratio of edge pixels that have come into the image to edge pixels that have disappeared. In a fade-in, there will be a strictly monotonic increase in newly appearing edge pixels; the opposite holds for a fade-out. A dissolve is comprised of a fade-in and a fade-out and can thus be recognized as their combination.

Wipes can be recognized by examining the local edge distribution and edge change. In a wipe, each image contains part of the preceding image and, at another location, part of the following image. Thus an isolated region of the image changes, while the other regions remain constant. During a horizontal wipe, there is a vertical strip that passes through the image either from right to left or left to right. In this strip, the edge change ratio is higher than in the edge regions. The problem is thus one of differentiating the strip region from other regions. Zabih [MMZ95] proposes a simple, yet robust method. The percentage of pixels that change is computed for the upper half of the image and for the left half of the image. During a horizontal wipe from left to right, most of the change will occur at the left at the beginning and at the right later. By computing change percentages for the four image quadrants at discrete times, it is possible to accurately determine the type of wipe. Moreover, in a wipe one does not find a characteristic ratio of newly appearing or disappearing edge pixels, so it is unlikely to be mistaken for a fade-in, a fade-out, or a dissolve. This is due in particular to the fact that the differences between both pixel change ratios are small since the change occurs only in a small area of the image.

9.3.2.6 Cut Detection through Chromatic Scaling

Hampapur *et al.* [Ham94, HJW94b, HJW95] describe an approach that identifies cut with the aid of the method described above, but uses a fundamentally different technique, chromatic scaling, to identify fade-ins, fade-outs, and dissolves.

The authors' cut detection method is based on cut boundaries and editing effects of partial scenes. A partial scene denotes a sequence of images that does not contain a cut. A scene can then be considered to be a set of partial scenes that belong together based on their content, for example, a dialog scene where the speakers are shot by two fixed cameras.

Hampapur *et al.* observe that editing images are never inserted between two images that form a cut boundary, though presumably they may be in the case of fade-ins or fade-outs and dissolves. They attempt to recognize these using chromatic scaling.

The basis of every video editing operation, with the exception of cuts, is the chromatic scaling of two superimposed video tracks. In a fade-in or fade-out, a black video track is used and the video track is scaled. In a dissolve, the two video tracks being mixed are both scaled [And88].

Commercial films primarily use two types of fades [Kuc87]: fade-in from black and fade-out to black. These can be modelled as chromatic scaling with positive and negative fade rates. A dissolve is the simultaneous scaling of two partial scenes and is thus the combination of a fade-in and a fade-out.

If one compares automatic transition detection methods, one finds that pixel-oriented methods recognize global features of the underlying images, whereas edge-oriented techniques tend to recognize local features. Local methods are more strongly affected by object movement than global methods, though they are more robust to brightness variations. The objective is thus to combine different approaches so as to achieve optimal transition detection. This can be accomplished as follows: since algorithms for detecting fade-ins, fade-outs, dissolves, and wipes are only known for edge-based approaches and for those based on chromatic scaling, these approaches can be used for detecting these effects. Some tests have shown that edge-based detection yields better results than chromatic scaling methods [Fis97b]. Both approaches yield similar results when recognizing fade-ins, fade-outs, and dissolves if a static image is scaled. Results with chromatic scaling are generally poorer for scenes in motion.

9.3.3 Analysis of Shots

In order to rapidly grasp the content of a shot, viewers often use the technique of "browsing." For video data, a representation is needed that greatly reduces the temporal resolution of the images. The goal is thus to find a few representative images for each shot. These so-called key frames are shown to stand in for the shot and give the viewer a rough idea of the shot's contents. When searching for specific shots in a database, this technique substantially reduces the amount of data that must be transmitted, since only the selected shots need to be made available in full.

Ideally, expressive images would be selected by finding points where interesting objects appear or interesting events occur. However, this requires semantic understanding, which until now is hardly ever possible except for special cases such as the detection of faces [LEJ98]. Even the selection of key frames generally uses low-level features and proceeds as follows:

S denotes the shot being analyzed, consisting of images $f_1, ..., f_n$, and R denotes the set of images selected as representative of S. R is initialized to $R = \{f_1\}$. Further, at any time f^* denotes for the last image added to R, and m denotes its index. Thus, initially $f^* = f_1$ and $m = 1$. The distance function $d_{feature}$ measures the distance between two images; ε is the maximum tolerable difference. If exceeded, a new key frame is taken. In any case, brief periods of interference or fluctuation of up to k images are disregarded [YL95].

```
do
    l  := min{l : l>m, l+k<=n, min{d_feature(f*,f_l),...,d_feature(f*,f_{l+k})}>ε}
    f* := f_l
    m  := l
    R  := R ∪ f*
while m<n-k
```

Examples of features that can be used to compare images include color or texture, or motion from image to image, since content change is frequently associated with camera motion or object movement [ZLSW95]. If multiple features are used, key frames should be chosen on a per feature basis. There are techniques for extracting key frames from sequences encoded in MPEG-1, MPEG-2, or motion JPEG that use features that can be computed directly from DCT coded images without requiring time-consuming decompression [DAM97].

9.3.4 Similarity-Based Search at the Shot Level

In the following it is assumed that a video is to be searched for shots that are similar to a given query shot. This problem arises, for example, if one is searching for the broadcast of a known commercial [Fis97b, RLE97]. Further we assume that feature vectors at the image level has already been computed as described in Section 9.2.

The first question that presents itself is how to represent shots to be compared. In practice, there are three approaches:

- Representation as an (unordered) set of key frames, which are described by feature vectors.
- Representation as an (ordered) sequence of key frames, which are likewise described by feature vectors.

- Representation as a disaggregated character string of feature vectors. For a shot with n images, the character string has n characters; each character is a feature vector.

Representation as a disaggregated set of images is not common.

9.3.4.1 Representation as a Set of Key Frames

This method assumes two shots S^1 and S^2 are available for which sets of representative key frames R^1 and R^2, respectively, have been chosen. The shot distance function is defined as follows, where a distance function $d_{feature}$ is used as a measure of the difference between two images [LEJ98]:

$$d_{shot}(S^1, S^2) = \sum_{k^1 \in R^1} \min\{d_{feature}(k^1, k^2) | k^2 \in R^2\}$$

The temporal ordering of the images is not considered here. A possible extension would be to compute mean values and deviation measures of some features as well as dominant colors across all images of the shot and to incorporate these into the comparison [ZLSW95].

9.3.4.2 Representation as a Sequence of Key Frames

If similarity also depends on whether or not the temporal ordering of the key frames in the shots being compared agrees, then the comparison algorithm presented in [DAM97] can be used. The sequence of key frames for shots S^1 and S^2 are denoted as $R^1 = \{k^1{}_1, ..., k^1{}_n\}$ and $R^2 = \{k^2{}_1, ..., k^2{}_m\}$, respectively, with $n < m$.

In the first step, $k^1{}_1$ is compared with $k^2{}_1$ and the distance is determined. Then $k^1{}_2$ is compared with the key frame(s) from R^2 that come next after $k^1{}_2$'s temporal position (i.e., its offset from the beginning of the corresponding shot), whereby at most a maximum difference is permitted. For example, if $k^1{}_2$ is the third image in S^1, then it will be compared with the second, third, and fourth images in S^2, provided that these are included in R^2. The distances (up to three) are added up and the number of addends is stored. If none of the images is contained in R^2, then the comparison is skipped for $k^1{}_2$. This process is repeated for all key frames in R^1, until finally the distance sum has been added up (Step 1 in Figure 9-11). The average is then computed by dividing by the number of addends.

By processing R^1 in this manner, R^1 is displaced relative to the beginning of S^2 such that the offset of $k^1{}_1$ corresponds to the offset of $k^2{}_2$ (Step 2 in Figure 9-11). R^1 is displaced by the next offset difference in S^2 a total $(m-n-1)$ times (the j-loop in the algorithm below). In each displacement iteration, the computed average distance is compared to the minimum of the previously computed average distances. If the new distance is smaller, it is stored as the new minimum and the current displacement is considered to be the optimal offset so far.

Analysis of Image Sequences

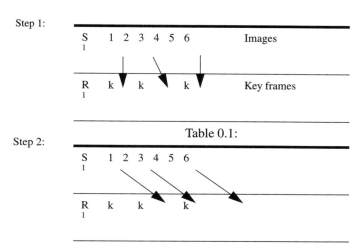

Figure 9-11 Identification of key frames.

The algorithm can be specified more precisely as follows: the function neighbor(R^2, $k^1{}_i$) computes the set of offsets of the images in R^2 whose respective offsets differ by at most one position from the offset of $k^1{}_i$; the number of elements in this set is thus between zero and three:

```
min := ∞
repeat for j := 1 to m-n
    sum_j := counter_j := 0
    repeat for i := 1 to n
        A := neighbor(R², k¹_i)
        sum_j +=  Σ_{a ∈ A} d_feature(k¹_i, k²_a)
        counter_j += |A|
    t := sum_j / counter_j
    if (t<min)
        min := t
        opt := j
    increase the offset of all images in R¹ by
    (offset of k²_{j+1} - offset of k²_j)
```

After the algorithm completes, `min` is checked to see if it less than a similarity threshold value. If so, then shots S^1 and S^2 are considered to be similar, and `opt` represents the offset of S^1 from the beginning of S^2 where the two shots best agree.

In order for the `neighbor` function to achieve an acceptable hit rate, the key frames in R^1 and R^2 must be densely packed (ε small). However, satisfying this requirement yields a smaller reduction in the amount of data, which was the goal of the key frames concept. This leads to the representation of a shot as a character string of feature vectors of all images.

9.3.4.3 Representation as Character String of Feature Vectors

In this approach, a shot S^2 is searched for a query shot S^1. The two shots are assigned feature vector character strings V^2 and V^1, respectively. The comparison is performed using an approximate string search for V^1 in V^2. The comparison is approximate in two respects. Firstly, when comparing two individual characters (feature vectors), they are only required to match approximately in ordered to be considered "equal." Secondly, V^1 is considered to be contained in V^2 even if a certain number of insertion, deletion, and character replacement operations are necessary in order to bring a substring of V^2 into agreement with V^1. Translated to the usual character alphabet, this means that "Muller" would be contained in "Hans_Müler-Schmidt," since "ü" and "u" match approximately, and the addition of an "l" to form "Muller" from "Müler" is allowed.

The algorithm used to perform the approximate string search must work efficiently for large alphabets (value range of the feature vectors). Appropriate algorithms are listed in [Ste94].

9.3.5 Similarity-Based Search at the Scene and Video Level

Analogous to the representation of a shot as a set or sequence of key frames or as a character string, one could represent a scene by using selected shots as well as by using all shots. Two scenes are then compared using methods analogous to those presented in Section 9.3.3 that work at a higher level of aggregation [LEJ98].

If one considers scenes in a video film, one can frequently recognize characteristic shot patterns according to which individual shots are grouped. In a dialog, there is an alternating series of shots from two classes (e.g., two interview partners shot in alternation). It is possible to directly search for such shot patterns, thereby allowing queries such as "find a dialog." In the following, shots that are similar to one another are indicated with the same letter of the alphabet, so that a dialog, for instance, can be visualized as "ABABABABA." A dialog is recognized by first searching for multiple occurrences of the same letter separated by short distances, whereby the distance between identical characters should be predominantly two. Once such a sequence is found, a test is performed to see if in the same manner, characters of a second class are arranged between the characters of the sequence. If such an embedded sequence is found, then the sequence represents a dialog. Note that the term "dialog" is used here in a broad sense—two parallel plot threads presented in alternation would also count as a dialog. If one is searching specifically for a conversation between two people, then additional search criteria must be provided.

In video production, it is not unusual to fall back on existing material that was used before either in a similar or in a completely different context. For example, one finds edited versions of original films or reports that incorporate archive material. Through a new cut, film material is used in a context that was not originally intended.

Two measures for comparing entire videos are proposed in [LEJ98]: a measure of the correspondence of two videos and a measure of the modification of the ordering of scenes within a single video (resequencing).

To compute these measures, video films are considered to be sequences of smaller units, for example, scenes or shots. Let $U^1 = \{u^1_1, ..., u^1_n\}$ denote the units that comprise video 1 and $U^2 = \{u^2_1, ..., u^2_m\}$ denote the units that comprise video 2. First the units in U^1 are compared with the units in U^2. Then an undirected graph is generated where the nodes represent the units and the edges connect units that are sufficiently similar to each other, as shown in Figure 9-12.

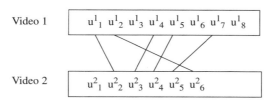

Figure 9-12 Similarity of videos.

If $|U^1 \cap U^2|$ denotes the number of units in U^1 for which there exists at least one sufficiently similar unit in U^2, then the correspondence measure is computed as follows:

$$\text{correspondence (video 1, video 2)} = \frac{|U^1 \cap U^2|}{|U_1|},$$

that is, the portion of the units in video 1 that are similar to units in video 2.

To test to what extent the common units in the videos occur in the same order, the resequencing measure is computed as follows:

```
numReOrd := numOrdAgree := 0
i₁ := i₂ := 1
while (i₂<m)
    i₂ := find the index of the first unit after i₂ in video 2
              from which there is an edge to video 1
    i_temp := find the index of the first unit after i₁ in video 1
              from which there is an edge to u²_{i₂}
    if (i_temp == 0)
      numReOrd++
      i₁ := find the first unit in video 1
              from which there is an edge to u²_{i₂}
    else
      numOrdAgree++
      i₁ := i_temp + 1
```

$$\text{ReSeq}(\text{video 1, video 2}) := \frac{\text{numReOrd}}{\text{numReOrd} + \text{numOrdAgree}}$$

A low value indicates that material that is common to the two videos appears in essentially the same order, from which one can conclude that context and content have been preserved. On the other hand, with a high resequencing value there is reason to suppose that due to reordering video material appears in a new context, and thus that possibly the content has been used in a way other than originally intended.

9.4 Audio Analysis

In the realm of digital audio, the syntactic indicators of volume and frequency distribution are available [BC94]. An audio signal can be fully defined through both of these indicators.

Furthermore, the following semantic indicators are conceivable, among others:

- speech fundamental frequency,
- fundamental frequency,
- onset, offset, and frequency transitions, and
- audio cut detection.

9.4.1 Syntactic Audio Indicators

The volume can be obtained from digital audio without any transformation if the data is available uncompressed or is decompressed before further processing.

The frequency distribution of an audio file is obtained by performing the Fourier transform. Although in the time domain there is a mixture of waves, their composition can be made visible in the frequency domain. Figure 9-13 shows a three-dimensional representation of the frequencies of a segment that contains silence followed by a short human exclamation. Time is on the horizontal axis, energy is on the vertical axis, and the frequency varies along the z-axis.

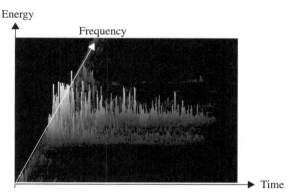

Figure 9-13 Frequency distribution of a sound mixture.

9.4.2 Semantic Audio Indicators

9.4.2.1 Fundamental Frequency

An important tool for analyzing the perception of complex sound mixtures is fundamental frequency or pitch. The fundamental frequency is the greatest common denominator of the frequencies where local maxima were observed [Bul77, Hes83, RDS76, GLCS95]. The following methods, among others, can be used to determine fundamental frequency [Fel85, GLCS95, Haw93, RJ93]: the cepstrum technique, the autocorrelation method, the maximum likelihood estimation method, and center clipping.

9.4.2.2 Audio Cut Detection

It is usually more difficult to implement cut detection in the audio domain than in the video domain [Fis94, Köh84]. The reason is that a video cut is easy to define as a series of two fundamentally different images. The audio domain, however, is different since a cut hierarchy can be specified. Examples of elements that can be used to define cuts in a digital audio include vowels in spoken language, words, sentences, and speech as opposed to other sound sources.

It is left to the user to set the decision threshold for recognizing a cut. This parameter must thus be treated as variable.

If one wants to roughly distinguish between speech and other sounds, it is possible to proceed according to the exclusion principle. Segments that have high energy values at frequencies above 7,000 Hz cannot be speech segments, since the portion of speech above 7,000 Hz is usually negligible.

If a sound mixture ranges only up to 7,000 Hz, it is not possible to make a determination. Nevertheless, the following simple technique can be used to implement cut detection in this case. Feature vectors consisting of volume, frequencies and speech fundamental frequency are computed for successive time windows i. The current value is then tested to see whether it is similar to the preceding value. An audio cut is detected if the distance measure between the vectors exceeds an established threshold value. The threshold can also be used to control the granularity of the cut detection. The technique described can also be used, for example to distinguish music from speech, even if both lie in the range up to 7,000 Hz. Among other things this is because the tone of music is not similar to that of speech.

Furthermore it is possible to extract "silence." However if noises occur after a quiet segment, the segment with the noises is much louder than the preceding segment. Thus to detect silence two conditions must be satisfied:

- The volume must be under a threshold value.
- The deviation between the segment and the preceding or following segment must be above a second threshold value.

9.4.2.3 Onset and Offset

Human auditory behavior is described as periodic increases or decreases in the response of auditory cells (measured via the cells' "firing probability"). Thus events that cause either a rise or drop in cell response are significant. These are measured using the two semantic indicators onset and offset. In the ear this mechanism is realized in the case of onset by applying a strong excitatory input followed by a strong inhibitory stimulus [WC72]. The inhibitory stimulus prevents the stimulus that is still applied from causing further cell reactions.

Both onset and offset can be set to more quickly identify signal changes that occur relatively slowly. In this context, a rapidly rising (falling) stimulus is referred to as fast onset (fast offset).

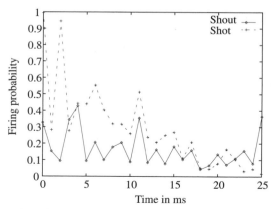

Figure 9-14 Onset of a shout and of a shot.

Figure 9-14 shows the application of a fast onset filter, which is useful for detecting particularly fast signal changes. Here one sees the fast onset of a shout and a shot. Using the fast onset indicator, the shout and the shot can be clearly distinguished [PFE96].

The calculation of the offset is the opposite of the calculation of the onset. Excitatory and inhibitory stimuli thus occur in the opposite order.

9.5 Applications

While the previous sections presented methods for automatic content analysis (with emphasis on "moving pictures"), the following sections present example applications of content analysis. These are based in part on directly derived syntactic indicators and in part on semantic indicators, such as scene length or fundamental frequency.

Content analysis can take place either through a classification or through a combination of various methods, such as recognizing actors by recognizing their voices. The applications listed below can be classified according to this scheme [Fis97b].

9.5.1 Genre Recognition

The term film genre is used to refer to a category or group of films that are similar with respect to the recording techniques used, subject, style, structure, or character. Genre categories are comprehensive enough that almost every film shot in a genre falls into a specific category. Nevertheless it is possible for some films to exist that have characteristics of several genres or that consist of several parts in different genres. Examples of genres are westerns, crime thrillers, soap operas or talk shows.

The basic principle behind automatic film genre recognition is to consider the computed indicators that form a feature vector to be like a fingerprint and to use a number of such fingerprints to create a database that can be used to recognize fingerprints of the same genre. To generate such a profile the data must be classified.

The following aspects must be considered in solving the classification problem:

- Based on performance and correlation a subset of the indicators available must be found that optimally solves the classification problem (data analysis).
- Quantitative weights must be determined for the individual indicators in the subset (transformation of the analysis results).
- Methods must be found to determine genre based on the weighted indicators (training of the system).
- The efficiency of the developed application must be specified quantitatively (evaluation through tests).

Fundamentally, indicators can be distinguished based on whether multiple values or only one value is available per time unit (image or time window in the audio domain). For the RGB color indicator, values are available for each pixel. On the other hand, for the image similarity indicator, there is a single value for each image pair.

9.5.1.1 Measuring the Performance of Indicators Used

First the performance of the indicators used must be analyzed and a subset needs to be determined that is always characteristic for the particular classification process. In view of their complexity and their quality, the preferred performance measures for this are nearest neighbor and inter- and intraclass distance.

The nearest neighbor performance measure ascertains the probability that the nearest neighbor of a point belongs to the feature space of the same class. The results thus lie in the range of zero to 100 percent.

The interclass distance performance measure specifies the average distance of the clusters of the feature vectors. A high interclass distance is desirable since the cluster features are homogeneous and easily separable. Unlike the nearest neighbor criterion, here distances can only be specified as absolute numbers. The evaluation of the indicators is then performed through a qualitative comparison of the absolute values.

The inter- and intraclass distance performance measure specifies the average distance of the clusters in combination with the average distance of the examples within a class. By adding the inverse intraclass distance, a high distance measure is also desirable in order to sufficiently separate the classes. As for the interclass distance, values can only be specified absolutely. The evaluation of the indicators is then performed through a qualitative comparison of the absolute values. Experiments have shown that a 1:2 weighting of the inter- to the intraclass distance produces the best results [Fis97b]. A higher weighting of the interclass distance results in features that have similar values for many examples of the classes being weighted heavier than features that have high values for the other criteria.

After evaluating the indicators using the different performance measures, the next step is to investigate how many indicators are necessary to describe a genre class.

9.5.1.2 Selection of Suboptimal Feature Sets

In the scope of selecting suboptimal feature sets, the following arise as central questions:

- Which quantization of the data sets should be selected?
- How many and which indicators should be selected in order to sufficiently describe the classes?

These questions are of central importance since most content analysis methods that include classification currently are far from real-time computations. If it is possible to specify a subset of the features that has performance criteria similar to the entire feature set, then it should definitely be done.

In principle, the efficiency of the performance measures presented can be compared according to the following criteria:

- the quality obtained,
- the required CPU time,
- the convergence rate of the sequential forward selection, wherein one indicator after another is added until the classification results appear satisfactory.

The quality obtained by the different methods cannot be compared directly, since each method has a different value range. The nearest neighbor performance measure specifies the percentage of the examples that belong to the same class, while the interclass distance and the inter- and intraclass distance performance gauges absolute distances between and within classes. A comparison of the values is thus not meaningful.

All the methods require approximately the same CPU time, so it is not possible to decide on a method based on this criterion.

The methods must therefore be judged according to the convergence of the sequential forward selection. One finds here that the nearest neighbor performance measure converges the fastest. A further advantage of this method is that the CPU time required is marginally less than for the other methods. Using this performance measure is thus advantageous.

9.5.1.3 Recognition Process

Classification for genre recognition can be performed in two steps:

- Computation of the indicators of a training set of films of different genres and building profiles for each genre.
- Comparison of films to be tested with all profiles via a Euclidean distance measure. The genre profile nearest to the test pattern is used to identify the recognized genre.

A profile represents a characterization of a class of examples. Thus to compute it a number of training examples are necessary from which the profile can be derived. Figure 9-15 shows the brightness for the genres news, comics, and music clips. One sees that these genres can be distinguished using the brightness indicator. A profile thus indicates the average distribution of the values of an indicator in a class of examples.

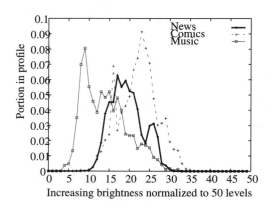

Figure 9-15 Brightness for genres newscast, comics, and music clips.

A Euclidean similarity measure can be used to compare the test example with the profiles. The profile that is most similar to the test example will be output as being recognized. As per the aforementioned considerations, the nearest neighbor performance measure should be used to select a suboptimal indicator set.

A problem with this approach is that test examples for which no profile exists could nevertheless be assigned to one of the profiles. This problem can be avoided by classifying examples as unknown unless the similarity measure exceeds a certain minimum value. Experiments have shown that requiring a similarity of at least 30 percent is sufficient [Fis97b].

The different recognition rates of the various classes of examples can be explained by the differing degrees of homogeneity of the classes. The music clip and commercial genres are less homogeneous than the newscast genre, which manifests itself in a worse recognition rate.

Class of Examples	Recognition Rate
Newscasts	91.3%
Music Clips	87.1%
Tennis	90.1%
Commercial	86.88%
Soccer	89.9%

Table 9-1 Recognition rates for the different genres.

9.5.2 Text Recognition in Videos

In television and movie theater films, newscasts or sports broadcasts, the superimposed text often supplies valuable supplementary information about the content of the video. Film credits list the names of the director and the actors, the place of the events are superimposed in news reports, and the current score appears in sports broadcasts. It thus stands to reason to search for such text occurrences when analyzing video sequences and to recognize them using the methods described in Section 9.3. A technique for automatically recognizing text in videos is described in [LS96].

Characters that were artificially added to a video sequence using a title machine—as opposed to text that happens to appear on objects—are distinguishable in that, among other things, they are monochrome and rigid, contrast with the background, have a certain minimum size and maximum size in order to be readable, and appear repeatedly in successive images while moving only linearly or not at all. These characteristics are taken as necessary prerequisites for extracting superimposed text in videos. This proceeds by incrementally identifying additional regions of an image as "nontext" and excluding them from subsequent processing until a suitable OCR algorithm can use the remaining regions as candidate characters.

First the original image is segmented using the split-and-merge technique (see Chapter 4) based on its gray values into homogeneous regions, so that eventually all characters are contained in one such region. Based on assumptions about the minimum and maximum sizes of characters, some of these regions can already be rejected.

In the next step, subsequent images are included in the analysis and motion analysis is used to identify corresponding character regions in successive images. For this a block matching method similar to the one used in MPEG-1 and MPEG-2 is used. To construct the blocks, the image is binarized (background black, everything else white) and each white pixel is temporarily expanded into a circle so that tightly packed characters are consolidated into a compact region. The delimiting rectangle of such a region is identified as a block. Regions in blocks that either do not appear in subsequent images or that appear with very different gray values are not considered further.

Then regions that have too low a contrast to the background are discarded.

The next step is to compute the ratio of the number of white pixels to the number of black pixels for each remaining block and to discard regions in blocks that have too low a fill factor.

The last segmentation step is to discard regions in blocks where the ratio of the width to the height does not lie within a preset interval.

For each image, one has kept only regions that have a high probability of being characters. For the hand-over to suitable OCR software, a new video is created in which for each image only the regions identified as candidate characters retain their original gray values and all other pixels are marked as background. The OCR procedure is applied to each image separately. Since each character generally appears in multiple successive images, possibly in a slightly changed position, an adjustment of the OCR results of these images is still necessary. The motion analysis described above is used for this, and it is possible to use the redundant information to simultaneously correct recognition errors that creep into individual images.

9.6 Closing Remarks

It has been shown that already a multitude of features exist that can be used to analyze video and audio. Reliable content analysis is thus not dependent on the development of new indicators, but on the skillful combination of these features to derive the semantic characteristics of digital films.

In the scope of this chapter, emphasis was laid on features for content analysis as well as on their possible combinations. It would also for example be conceivable to couple the methods presented here with the results of pattern recognition analysis. In the process of genre recognition, for example, if a video is classified as a newscast, it would be possible to additionally search for a logo or a newscaster if the analysis results do not permit a unique demarcation to a further genre.

This chapter emphasized analysis of the contents of locally available data. It would also be possible to use the knowledge gained here for transmitting and storing data, for example, to support content-oriented compression, such as MPEG-4, or for content-oriented scaling of data during network transmission.

For further literature, the reader is referred to [Jäh97] and to [AZP96] for a highly readable survey article.

Bibliography

[Ace93]   A. Acero. *Acoustical and Environmental Robustness in Automatic Speech Recognition*. Kluwer Academic Publishers, Boston, 1993.

[ACG93]   B. S. Atal, V. Cuperman, and A. Gersho, editors. *Speech and Audio Coding for Wireless and Network Applications*. Kluwer Academic Publisher, Dortrecht, 1993.

[ACM89]   *Communications of the ACM: Special Section on Interactive Technology*, July 1989.

[ADHC94] F. Arman, R. Depommier, A. Hsu, and M.-Y. Chiu. "Content-based browsing of video sequences." In *Proceedings of Second ACM International Conference on Multimedia*, pages 97–103, Anaheim, CA, October 1994.

[AFN90]   Y. Ashikaga, K. Fukuoka, and M. Naitoh. "CD-ROM Premastering System Using CD-Write Once." *Fujitsu Scientific Technical Journal*, 26(3):214–223, October 1990.

[AGH90]   D.P. Anderson, R. Govindan, and G. Homsy. "Abstractions for Continuous Media in a Network Window System." Technical Report UCB/CSD 90/596, Computer Science Division, UC Berkeley, Berkeley, CA, September 1990.

[AHC93]   F. Arman, A. Hsu, and M.-Y. Chiu. "Image processing on compressed data for large video databases." *ACM Multimedia*, pages 267–272, June 1993.

[AJ94] P. Aigrain and P. Joly. "The automatic real-time analysis of film editing and transition effects and its applications." *Computers & Graphics*, 18(1):93–103, January–February 1994.

[All85] J. B. Allen. "Cochlear modelling." *IEEE ASSP Magazine*, pages 3–29, 1985.

[Ana87] P. Anandan. *Measuring visual motion from image sequences*. PhD thesis, Dept. of Computer Science, University of Massachussetts, Amherst, MA, 1987.

[Ana89] P. Anandan. "A computational framework and an algorithm for the measurement of visual motion." *International Journal Computer Vision*, 2:283–310, 1989.

[And88] H.G. Anderson. *Video Editing and Post Production: A Professional Guide*. Knowledge Industry Publications, 1988.

[ANM96] Ames, Nadeau, and Moreland. *VRML 2.0 Sourcebook*. Wiley and Sons Inc., NY, 1996.

[Ann94a] Announcement. "3D Digitizing Systems." *Computer Graphics World*, page 59, April 1994.

[Ann94b] Announcement. "Kodak Expands Photo CD System." *Computer Graphics World*, page 59, April 1994.

[Ann94c] Announcement. "Video Accelerator Boards." *Computer Graphics World*, page 53, May 1994.

[ANR74] N. Ahmed, T. Natarajan, and K. R. Rao. "Discrete Cosine Transform." *IEEE Transactions on Computers*, 23:90–93, January 1974.

[AZP96] P. Aigrain, H. Zhang, and D. Petkovic. "Content-based representation and retrieval of visual media: A state-of-the-art review." *Multimedia Tools and Applications*, 3:179–202, 1996.

[Bae69] R. M. Baecker. "Picture Driven Animation." In *Proceedings of Spring Joint Computer Conference*, pages 273–288, Montvale, NJ, 1969. AFIPS Press.

[Bas90] G. A. J. Bastiaens. "Compact Disc Interactive." *Computer Education*, pages 2–5, February 1990.

[BC94] G.J. Brown and M. Cooke. "Computational auditory scene analysis." *Computer Speech and Language*, 8:297–336, August 1994.

[BD97] L. Brun and J.P. Domenger. "A new split and merge algorithm based on discrete map." In N.M. Thalmann and V. Skala, editors, *The Fifth International Conference in Central Europe on Computer Graphics and Visualization (WSCG '97)*, pages 21–30, Plzen-Bory, Chech Republic, 1997.

[Beg94] D. R. Begault. *3-D Sound for Virtual Reality and Multimedia*. Academic Press, London, 1994.

[BF91] K. B. Benson and D. G. Fink. *HDTV—Advanced Television for the 1990s*. Intertext Publications, McGraw-Hill Publishing Company, Inc., 1991.

[BFB94] J.L. Barron, D.J. Fleet, and S.S. Beauchemin. "Performance of optical flow techniques." *International Journal of Computer Vision*, 12(1):43–77, January 1994.

[BH93] M. F. Barnsley and L. P. Hurd. *Fractal Image Compression*. AK Peters, Ltd., Wellesley, Massachusetts, 1993.

[BHS91] G. Blair, D. Hutchison, and D. Shepard. "Multimedia systems." In *Tutorial Proceedings of 3rd IFIP Conference on High-Speed Networking*, Berlin, Germany, March 18–22, 1991.

[Bla71] J. Blauert. "Localization and the law of the first wavefront in the median plane." *Journal of the Acoustical Society of America*, 50:466–470, 1971.

[Bla74] J. Blauert. *Räumliches Hören*. S. Hirzel Verlag, Stuttgart, 1974.

[BN93] Ch. Baber and J. M. Noyes, editors. *Interactive Speech Technology: Human Factors Issues in the Application of Speech Input/Output to Computers*. Taylor & Francis, Bristol, PA, 1993.

[Boo87] M. Boom. *Music Through MIDI*. Microsoft Press, 1987.

[Bor79] A. Borning. "Thinglab—A Constraint-Oriented Simulation Laboratory." Technical Report SSI-79-3, Xerox Palo Alto Research Center, Palo Alto, CA, July 1979.

[Bow92] Sing-Tze Bow. *Pattern Recognition and Image Preprocessing*. Marcel Dekker, Inc., New York, NY, 1992.

[Bra87] S. Brand. *The Media Lab, Inventing the Future at MIT*. Viking Penguin, New York, 1987.

[Bri86] G. Bristow, editor. *Electronic Speech Recognition: Techniques, Technology, and Applications*. McGraw-Hill Publishing Company, Inc., New York, 1986.

[BS83]   W. Becker and N. Schöll. *Methoden und Praxis der Filmanalyse*. Leske und Budrich Verlag, Opladen, 1983.

[Bul77]   T.H. Bullock, editor. *Recognition of Complex Acoustic Signals*, Volume 5 of *Life Sciences Research Report*. Abakon Verlagsgesellschaft, Berlin, 1977.

[B.V89]   Philips International B.V. *Compact Disc-Interactive—A Designers Overview*. Kluwen, Technische Boeken B.V., Deventer, Netherlands, 1989.

[BW76]   N. Burtnyk and M. Wein. "Interactive Skeleton Techniques for Enhancing Motion Dynamics in Key Frame Animation." *Communications of the ACM*, 19(10):564–569, October 1976.

[BW90a]   C. Biaesch-Wiebke. *Compact Disc Interactive*. Vogel Buchverlag Würzburg, Vogelfachbuch: Kommunikationstechnik, 1990.

[BW90b]   C. Biaschk-Wiebke. *Compact Disc Interactive*. Vogel-Verlag, Würzburg, 1990.

[Can86]   J. Canny. "A computational approach to edge detection." In *IEEE Transactions on Pattern Analysis and Machine Intelligence, Volume PAMI-VIII, No. 6*, November 1986.

[CCI82]   International Radio Consultative Committee (CCIR). *Recommendation 601: Encoding Parameters of Digital Television for Studios, vol. 11, Teil 1, pp. 271-273*. CCIR, Geneva, 1982.

[Cho71]   J. Chowning. "The simulation of a moving sound source." *Journal of the Audio Engineering Society*, 41(11), 1971.

[CLP94]   T.-S. Chua, S.-K. Lim, and H.-K. Pung. "Content-based retrieval of segmented images." In *ACM Multimedia '94*, pages 211–218, Anaheim, CA, October 1994.

[CMVM86]   F. Cheevasuvut, H. Maitre, and D. Vidal-Madjar. "A robust method for picture segmentation based on a split and merge procedure." In *Computer Vision, Graphics, and Image Processing*, 34:268–281, 1986.

[Cor92]   Aldus Corporation. *TIFF—Tagged Image File Format—Revision 6.0 Final*. Aldus Corporation, Seattle, 1992.

[CP79]   P. Chen and T. Pavlidis. "Segmentation by texture using a co-occurence matrix and a split and merge algorithm." In *Computer Graphics and Image Processing*, pages 172–182, 1979.

[DAM97]   N. Dimitrova and M. Abdel-Mottaleb. "Content-based video retrieval by example video clip." In *Storage and Retrieval for Image and Video Databases (SPIE), SPIE Proceedings Vol. 3022*, pages 59–70, San Jose, CA, 1997.

[DB81]   U. Dehm and G. Bentele, editors. *Thesen zum Vergleich elektronischer und konventioneller Inhaltsanalyse*, chapter Semiotik und Massenmedien. Ölschläger Verlag, München, 1981.

[Dep89]   Marketing Department. Brochure. Showscan Film Corporation, 1989.

[DG82]   B.I. Slikvood, D. Goedhart, and R. J. van de Plassche. "Digital to analog conversion in playing a compact disc." *Philips Technical Review*, 40(6), August 1982.

[DG90]   P. Duhamel and C. Guillemot. "Polynomial transform computation of the 2-D DCT." In *Proceedings of IEEE ICASSP-90*, pages 1515–1518, Albuquerque, New Mexico, 1990.

[ECM88]   ECMA. *Data Interchange on Read-Only 120mm Optical Data Disks (CD-ROM)*. European Computer Manufacturers Association Standard ECMA-130, 1988.

[EF94]   J.J. Encarnacao and J.D. Foley. *Multimedia*. Springer-Verlag, Berlin, 1994.

[End84]   W. Endres. "Verfahren zur Sprachsynthese—Ein geschichtlicher Überblick." *Der Fernmelde-Ingenieur*, pages 2–40, September 1984.

[ES98]   W. Effelsberg and R. Steinmetz. *Video Compression Techniques*. dpunkt-Verlag, Heidelberg, 1998.

[Eur94]   European Telecommunications Standards Institute. *Digital Broadcasting Systems for Television, Sound and Data Services: Specification for Service Information (SI) in Digital Video Broadcasting Systems*. ETSI, ETSI 300 468 edition, November 1994.

[Eur96]   European Telecommunications Standards Institute. *Digital Broadcasting Systems for Television, Sound and Data Services: Framing, Structure, Channel Coding and Modulation for Digital Terrestrial Television*. ETSI, ETSI 300 744 edition, April 1996.

[Fal85]   F. Fallside. *Computer Speech Processing*. Englewood Cliffs, NJ: Prentice-Hall International, 1985.

[FDFH92]   J. D. Foley, A. van Dam, S. K. Feiner, and J. F. Hughes. *Computer Graphics—Principles and Practice*. Addison-Wesley Publishing Company, Inc., 2nd edition, 1992.

[FE88]   E. A. Fox and M. E. Williams, editors. *Optical disks and CD-ROM: Publishing and Access*. Elsevier Science Publishers, 1988.

[Fei90]   E. Feig. "A fast scaled DCT algorithm." In K. S. Pennington and R. J. Moorhead II, editors, *Image Processing Algorithms and Techniques*, volume 1244, pages 2–13, Santa Clara, CA, February 11–16, 1990.

[Fel85]   K. Fellbaum. *Sprachverarbeitung und Sprachsynthese*. Springer-Verlag, Berlin, 1985.

[Fis94]   Alon Fishbach. "Primary segmentation of auditory scenes." In *International Conference on Pattern Recognition (ICPR)*, pages 113–117, Jerusalem, Israel, 1994.

[Fis97a]   S. Fischer. "Image segmentation by water-inflow." In *Proceedings of WSCG97, The Fifth International Conference in Central Europe on Computer Graphics and Visualization*, pages 134–143, Plzen-Bory, Czech Republic, 1997.

[Fis97b]   S. Fischer. *Indikatorenkombination zur Inhaltsanalyse digitaler Filme*. PhD thesis, Universität Mannheim, 1997.

[FJ90]   D.J. Fleet and A.D. Jepson. "Computation of component image velocity from local phase information." *International Journal of Computer Vision*, 5:77–104, 1990.

[FKT96]   Fkt—Fachzeitschrift für Fernsehen, Film und Elektronische Medien. Schwerpunkt: Digital Video Broadcasting. April 1996.

[Fla72]   J. L. Flanagan. *Speech Analysis, Synthesis and Perception*. Springer-Verlag, New York, 1972.

[Fla92]   J. L. Flanagan. "Speech technology and computing: A unique partnership." *IEEE Communications Magazine*, pages 84–89, May 1992.

[FLE95]   S. Fischer, R. Lienhart, and W. Effelsberg. "Automatic recognition of film genres." In *Proceedings of Third ACM International Conference on Multimedia*, pages 295–304, San Francisco, CA, November 1995.

[Fri92a]   J. R. Frick. "Compact Disc Technology." Internal Publication, 1992. Disc Manufacturing, Inc.

[Fri92b] G. Fries. Lautspezifische Synthese von Sprache im Zeit- und im Frequenzbereich. *VDI Fortschritts-Berichte*, 10(213), 1992. VDI Verlag.

[FS92] S. Furui and M. M. Sondhi, editors. *Advances in Speech Signal Processing*. Marcel Dekker Inc., New York, Hong Kong, 1992.

[FSB82] S. Feiner, D. Salesin, and T. Banchoff. "DIAL: A diagrammatic animation language." In *Computer Graphics and Applications*, 2(7):43–54, September 1982.

[Fur98] Borko Furht, editor. *Handbook of Multimedia Computing*. CRC Press, Boca Raton, Florida, 1998.

[GC89] P. Ghislandi and A. Campana. "In Touch with XA. Some Considerations on Earlier Experiences of CD-ROM XA Production." In *Proceedings of 13th International Online Information Meeting*, pages 211–226, London, 1989.

[Gha99] Mohammed Ghanbari. *Video Coding: An Introduction to Standard Codecs*. Number 42 in Telecommunications Series. IEE Press, London, UK, 1999.

[GLCS95] A. Ghias, J. Logan, D. Chamberlain, and B.C. Smith. "Query by humming: Musical information retrieval in an audio database." In *Proceedings of Third ACM International Conference on Multimedia*, pages 231–236, San Francisco, CA, November 1995.

[GM94] W. G. Gradner and K. Martin. "HRTF Measurements of a KEMAR Dummy." Technical Report 280, MIT Media Lab Perceptual Computing, 1994.

[Gra84] R. M. Gray. "Vector quantization." *IEEE Acoustics, Speech, and Signal Processing Magazine*, 1(2):4–29, April 1984.

[GV92] C. Gonzales and E. Viscito. Flexible Digital Video Coding. Personal Communication, 1992.

[GW93] R. C. Gonzales and R. E. Woods. *Digital Image Processing*. Addison Wesley Publishing Company, Inc., 1993.

[GWJ92] A. Gupta, T. Weymouth, and R. Jain. "Semantic queries in image databases." In E. Knuth and L.M.Wegner, editors, *IFIP Transactions A: Computer Science and Technologies*, Volume A-7, pages 201–215, Amsterdam, Netherlands, 1992.

[Hab95] P. Haberäcker. *Praxis der digitalen Bildverarbeitung und Mustererkennung*. Hanser Studienbücher, Carl Hanser Verlag, München, Wien, 1995.

[Ham94]   A. Hampapur. *Designing Video Data Management Systems*. PhD thesis, University of Michigan, 1994.

[Hat82]   Hatada. Psychophysical experiments on a widefield display. Technical Report 276, NHK, July 1982.

[Haw93]   M.J. Hawley. *Structure out of sound*. PhD thesis, Massachussetts Institute of Technology, MA, September 1993.

[HB90]   J.F. Haddon and J.F. Boyce. "Image segmentation by unifying region and boundary information." *IEEE Transactions on Pattern Analysis and Machine Intelligence*, PAMI-12(10):929–948, October 1990.

[HD90]   R. G. Herrtwich and L. Delgrossi. "ODA-based data modeling in multimedia systems." Technical Report 90-043, International Computer Science Institute, Berkeley, 1990.

[Hee87]   D.J. Heeger. "Model for the extraction of image flow." *Journal of the Optical Society of America*, Am A4:1455–1471, 1987.

[Hee88]   D.J. Heeger. "Optical flow using spatiotemporal filters." *International Journal on Computer Vision*, 1:279–302, 1988.

[Hen95]   R.D. Henkel. "Segmentation in scale space." In *6th International Conference on Computer Analysis of Images and Pattern (CAIP)*, 1995.

[Hes83]   W. Hess. *Pitch determination of Speech Signals*. Springer-Verlag, Berlin, Heidelberg, 1983.

[HHS96]   T. Hong, J.J. Hull, and S. N. Srihari. "A unified approach towards text recognition." In *IS&T/SPIE's Symposium on Electronic Imaging: Science & Technology*, San Jose, CA, January 1996.

[HJW94a]   A. Hampapur, R. Jain, and T. Weymouth. "Digital video indexing in multimedia systems." In *Proceedings of the Workshop on Indexing and Reuse in Multimedia Systems*. American Association of Artificial Intelligence, Aug. 1994.

[HJW94b]   A. Hampapur, R. Jain, and T. Weymouth. "Digital video segmentation." In *Proceedings of Second ACM International Conference on Multimedia*, pages 357–364, San Francisco, CA, October 1994.

[HJW95]   A. Hampapur, R. Jain, and T.E. Weymouth. "Production model based digital video segmentation." *Multimedia Tools and Applications*, 1:9–46, March 1995.

[HKL+91]   K. Harney, M. Keith, G. Lavelle, L. D. Ryan, and D. J. Stark. "The i750 Video Processor: A total multimedia solution." *Communications of the ACM*, 34(4):64–78, April 1991.

[HKR93]   D. Huttenlocher, G. Klanderman, and W. Rucklidge. "Comparing images using the hausdorff distance." *IEEE Transactions PAMI*, 15(9):850–863, 1993.

[Hol88]   F. Holtz. *CD-ROM: Breakthrough in Information Storage*. TAB Books, Inc., 1988. ISBN 0-8306-1426-5.

[Hou88]   H. S. Hou. "A fast recursive algorithm for computing the discrete cosine transform." *IEEE Transactions on Acoustics, Speech and Signal Processing*, ASSP-35(10):1455–1461, 1988.

[HS81]   B. Horn and B. Schunck. Determining optical flow. *Artificial Intelligence*, 17:185–203, 1981.

[HS82]   J. P. J. Heemskerk and K. A. Schouhamer Immink. "Compact disc: System aspects and modulation." *Philips Technical Review*, 40(6), August 1982.

[HS91]   R. G. Herrtwich and R. Steinmetz. "Towards integrated multimedia systems: Why and how." *Informatik-Fachberichte*, 293:327–342, 1991. Springer-Verlag.

[HS92]   R. M. Haralick and L. G. Shapiro. *Computer and Robot Vision*, Volume 1. Addison-Wesley Publishing Company, Inc., 1992.

[HTV82]   H. Hoeve, J. Timmermas, and L. B. Vries. "Error correction and concealment in the compact disc system." *Philips Technical Review*, 40(6), August 1982.

[Huf52]   D. A. Huffman. "A method for the construction of minimum redundancy codes." In *Proceedings of IRE 40*, pages 1098–1101, September 1952.

[HYS88]   G. Hudson, H. Yasuda, and I. Sebestyen. "The international standardization of a still picture compression technique." In *Proceedings of IEEE Global Telecommunications Conference*, pages 1016–1021, November 1988.

[Inc85]   Symbolics Inc. *S-Dynamics*. Symbolics, Inc., Cambridge, MA, 1985.

[Inc90]   Philips International B. V. Adobe Systems Incorporated. *Adobe Type 1 Format*. Adobe Systems Incorporated, 1990.

[Ins94]　　European Telecommunications Standards Institute. *Digital Broadcasting Systems for Television; Implementation Guidelines for the Use of MPEG-2 Systems, Video and Audio in Satellite and Cable Broadcasting Applications*. ETSI, ETSI Technical Report 154 edition, November 1994.

[Ins95]　　European Telecommunications Standards Institute. *Digital Broadcasting Systems for Television, Sound and Data Services; Framing, Structure, Channel Coding and Modulation for CATV Cable and SMATV Distribution*. ETSI, ETSI 300 473 edition, April 1995.

[ISO93a]　ISO. Information Technology—Coding of Moving Pictures and Associated Audio for Digital Storage Media, Test Model 4. Draft, MPEG 93/255b, February 1993. ISO IEC JTC 1.

[ISO93b]　ISO. Information Technology—Coding of Moving Pictures and Associated Audio for Digital Storage Media up to about 1.5 Mbit/s, 1993. ISO IEC JTC1/SC29.

[ITUC90]　The International Telegraph International Telecommunication Union and Telephone Consultative Committee. Line Transmission on non-Telephone Signals: Video Codec for Audiovisual Services at $p \times 64$ kbit/s. CCITT Recommendation H.261, 1990.

[Jäh97]　　B. Jähne. *Digitale Bildverarbeitung*. Springer-Verlag, Berlin, 4th edition, 1997.

[Jai89]　　A. K. Jain. *Fundamentals of Digital Image Processing*. Prentice Hall, Inc., Englewood Cliffs, New Jersey, 1989.

[JLW95]　J. M. Jot, V. Lachner, and O. Warusfel. *Digital Signal Processing Issues in the Context of Binaural and Transaural Stereophony*. Audio Engineering Society, 1995. Reprint 3980.

[JN84]　　N. S. Jayant and Peter Noll. *Digital Coding of Waveforms*. Prentice-Hall, 1984.

[Joh92]　　J. Johann. *Modulationsverfahren*. Springer-Verlag, 1992. Reihe Nachrichtentechnik.

[KDS95]　M. Kleiner, B. Dalenbäck, and P. Svenson. "Auralization—an overview." *Journal of the Audio Engineering Society*, 41(11), 1995.

[Ken95]　　G. S. Kendall. "A 3-D sound primer: Directional hearing and stereo reproduction." *Computer Music*, 19(4), 1995.

[KGTM90] P.H. Kao, W. A. Gates, B. A. Thompson, and D. K. McCluskey. "Support for the ISO 9669/HSG CD-ROM File System Standard in the HP-UX Operating System." *Hewlett-Packard Journal*, pages 54–59, December 1990.

[KJ91] R. Kasturi and R. Jain, editors. *Dynamic vision*. IEEE Computer Society Press, 1991.

[KJB96] K. Karu, A. K. Jain, and R. M. Bolle. "Is there any texture in the image?" *Pattern Recognition*, 29(9):1437–1446, 1996.

[Kle92] B. Klee. "CD-ROM/WO also Kompatibles Publishing Medium." In *Proceedings of DGB-Online Tagung*, Frankfurt, Germany, April 1992.

[Köh84] M. Köhlmann. *Rhythmic Segmentation of Sound Signals and Their Application to the Analysis of Speech and Music (Rhythmische Segmentierung von Schallsignalen und ihre Anwendung auf die Analyse von Sprache und Musik)*. PhD thesis, Technische Universität München, 1984. (in German).

[KR82] A. C. Kak and A. Rosenfeld. *Digital Picture Processing*, Volume 1. Academic Press, 2nd edition, 1982.

[KSN+87] S. Komatsu, T. Sampel, T. Nishihara, T. Furuuya, and Y. Yamada. "The multimedia CD-ROM system for educational use." *IEEE Transactions on Consumer Electronics*, 33(4):531–539, November 1987.

[Kuc87] T. Kuchenbuch. *Filmanalyse*. Prometh Verlag, Köln, 1987.

[KW93] B. Klauer and K. Waldschmidt. "An object-oriented character recognition engine." In *Proceedings of European Informatik Congress, Euro-ARCH'93*. Springer-Verlag, October 1993.

[Lan84] G. Langdon. "An Introduction to Arithmetic Coding." *IBM Journal of Research and Development*, 28:135–149, March 1984.

[LCP90] S. Lee, S. Chung, and R. Park. "A comparative performance study of several global thresholding techniques for segmentation." In *Computer Vision, Graphics and Image Processing*, 52:171–190, 1990.

[Le 91] D. Le Gall. "MPEG: A video compression standard for multimedia applications." *Communications of the ACM*, 34(4):46–58, April 1991.

[LE91] B. Lamparter and W. Effelsberg. "X-MOVIE: Transmission and presentation of digital movies under X." In *Proceedings of 2nd International Workshop on Network and Operating System Support for Digital Audio and Video*, pages 18–19, Heidelberg, Germany, November 1991.

[Lee84] B. G. Lee. "A new algorithm to compute the discrete cosine transform." *IEEE Transactions on Acoustic Speech and Signal Processing*, ASSP–32(6):1243–1245, December 1984.

[LEJ98] R. Lienhart, W. Effelsberg, and R. Jain. "Towards a visual grep: A systematic analysis of various methods to compare video sequences." In K. Sethi and Ramesh C. Jain, editors, *Storage and Retrieval for Image and Video Databases VI, Proc. SPIE 3312*, pages 271–282, 1998.

[LEM92] B. Lamparter, W. Effelsberg, and N. Michl. "MTP: A movie transmission protocol for multimedia applications." In *Proceedings of the 4th IEEE ComSoc International Workshop on Multimedia Communications*, pages 260–270, Monterey, CA, April 1992.

[LF91] E. N. Linzer and E. Feig. "New DCT and scaled DCT algorithms for fused multiply/add architectures." In *Proceedings of IEEE International Conference on Acoustics, Speech, and Signal Processing (ICASSP)*, pages 2201–2204, Toronto, Canada, May 1991.

[Lio91] M. Liou. "An overview of the $p \times 64$ kbit/s video coding standard." *Communications of the ACM*, 34(4):59–63, April 1991.

[Lip91] A. Lippman. "Feature sets for interactive images." *Communications of the ACM*, 34(4):92–101, April 1991.

[LK81] B. Lucas and T. Kanade. "An iterative image registration technique with an application to stereo vision." In *Department of Defence Advanced Research Project Agency (DARPA) Image Understanding Workshop*, pages 121–130, 1981.

[LMY88] A. Leger, J. Mitchell, and Y. Yamazaki. "Still picture compression algorithm evaluated for international standardization." In *Proceedings of IEEE Global Telecommunications Conference*, pages 1028–1032, November 1988.

[LOW91] A. Leger, T. Omachi, and G. K. Wallace. "JPEG still picture compression algorithm." *Optical Engineering*, 30(7):947–954, July 1991.

[Loy85] C. Loy. "Musicians make a standard: The MIDI phenomenon." *Computer Music Journal*, 9(4), 1985.

[LPE97] R. Lienhart, S. Pfeiffer, and W. Effelsberg. "Video abstracting." *Communications of the ACM*, 40(12):55–62, 1997.

[LR86] S. Lambert and S. Roplequet. *CD-ROM: The New Papyrus*. Microsoft Press, Redmond, WA, 1986.

[LS96] R. Lienhart and F. Stuber. "Automatic text recognition in digital videos." In *Image and Video Processing IV, Proc. SPIE 2666-20*, 1996.

[Luc84] B.D. Lucas. *Generalized Image Matching by the Method of Differences*. PhD thesis, Dept. of Computer Science, Carnegie-Mellon University, Pittsburgh, 1984.

[Lut91] A. C. Luther. *Digital Video in the PC Environment*. Intertext Publications McGraw-Hill Publishing Company, Inc., New York, 1991.

[Lut94] A. C. Luther. *Authoring Interactive Multimedia*. Academic Press, 1994.

[LWT94] C. J. Lindblad, D. J. Wetherall, and D. L. Tennenhouse. "The VuSystem: A programming system for visual processing of digital video." In *Proceedings of Second ACM International Conference on Multimedia*, pages 307–314, Anaheim, CA, October 1994.

[Mam93] R. J. Mammone, editor. *Artificial Neural Networks for Speech and Vision*. Chapman & Hall, London, New York, 1993.

[Mei83] B. Meier. BRIM. Technical report, Computer Graphics Group, Computer Science Department, Brown University, Providence, RI, 1983.

[Mey91] K. Meyer-Wegener. *Multimedia Datenbanken*. B. G. Teubner, Stuttgart, 1991.

[MGC82] J. P. Sinjou, M. G. Carasso, and J. B. H. Peek. "The compact disc audio system." *Philips Technical Review*, 40(6), August 1982.

[MHE93] MHEG. *Information Technology—Coded Representation of Multimedia and Hypermedia Information (MHEG), Part 1: Base Notation (ASN.1)*. Committee draft ISO/IEC CD 13522-1, June 1993. ISO/IEC JTC1/SC29/WG12.

[MMZ95] K. Mai, J. Miller, and R. Zabih. "A feature-based algorithm for detecting and classifying scene breaks." In *Proceedings of Third ACM International Conference on Multimedia*, pages 189–200, San Francisco, CA, November 1995.

[Moo79] James A. Moorer. "About this reverberation business." *Computer Music*, 3(2):13–28, 1979.

[MP91] J. L. Mitchell and W. B. Pennebaker. "Evolving JPEG color data compression standard." In M. Nier and M. E. Courtot, editors, *Standards for Electronic Imaging Systems*, Volume CR37, pages 68–97. SPIE, 1991.

[Mus90] H. G. Musmann. "The ISO audio coding standard." In *IEEE Globecom 90*, pages 511–517, San Diego, CA, December 1990.

[Nag83] H.H. Nagel. "Displacement vectors derived from second-order intensity variations in image sequences." *Computer Graphics and Image Processing*, 21:85–117, 1983.

[Nag89] H.H. Nagel. "On a constraint equation for the estimation of displacement rates in image sequences." *IEEE Transactions on Pattern Analysis and Machine Intelligence*, 11:13–30, 1989.

[Nev82] R. Nevatia. *Machine Perception*. Prentice-Hall, Inc., Englewood Cliffs, NJ, 1982.

[NH88] A. N. Netravali and B. G. Haskell. *Digital Pictures: Representation and Compression*. Plenum Press, New York, 1988.

[NP78] N. J. Narasinha and A. M. Peterson. "On the computation of the discrete cosine transform." *IEEE Transactions on Communications*, COM-26(6):966–968, October 1978.

[NT91] A. Nagasaka and Y. Tanaka. "Automatic video indexing and full-video search for object appearances." In *Proceedings of the Second Working Conference Visual Database Systems*, pages 119–133, 1991.

[O'86] D. O'Shaughnessy. "Speaker recognition." *IEEE Acoustics, Speech, and Signal Processing Magazine*, 3(4):4–17, October 1986.

[OC89] S. Oberlin and J. Cox, editors. *Microsoft CD-ROM Yearbook 1989–1990*. Microsoft Press, Redmond, WA, 1989.

[OO91] M. Ohta and S. Ono. "Super high definition image communication—application and technologies." In *Fourth International Workshop on HDTV and Beyond*, September 1991.

[Org93] International Standards Organization. Information Technology—Digital Compression and Coding of Continuous-Tone Still Images. International Standard ISO/IEC IS 10918, 1993. ISO IEC JTC 1.

[Org96] International Standards Organization. Information Technology—Generic Coding of Moving Pictures and Associated Audio Information (MPEG-2). International Standard ISO/IEC IS 13818, 1996. ISO IEC 1.

[Org97] International Standard Organization. Information Technology—Computer Graphics and Image Processing—Virtual Reality Modeling Language (VRML). ISO/IEC, Geneva, 1997. ISO 14772.

[O'S90]   D. O'Shaughnessy. *Speech Communication*. Addison-Wesley Publishing Company, Inc., Reading, Massachusetts, 1990.

[OT93]   K. Otsuji and Y. Tonomura. "Projection detection filter for video cut detection." In *Proceedings of ACM Multimedia 93 (Anaheim, CA, USA, August 1-6, 1993)*, pages 251–257. ACM, New York, 1993.

[PA91]   A. Puri and R. Aravind. "Motion compensated video coding with adaptive perceptual quantization." *IEEE Transactions on Circuits and Systems for Video Technology*, 1:351, December 1991.

[Per97]   J. Perl, editor. *Informatik im Sport*. Verlag Karl Hofmann, Schorndorf, 1997.

[PFE96]   S. Pfeiffer, S. Fischer, and W. Effelsberg. "Automatic audio content analysis." In *Proceedings of Fourth ACM International Conference on Multimedia*, pages 21–30, November 1996.

[Phi73]   Phillips. Laser vision. *Phillips Technical Review*, 33:187–193, 1973.

[Phi82]   Phillips and Sony Corporation. *System Description Compact Disc Digital Audio*, 1982. Red Book.

[Phi85]   Phillips and Sony Corporation. *System Description Compact Disc Read Only Memory*, 1985. Yellow Book.

[Phi88]   Phillips and Sony Corporation. *CD-I Full Functional Specification*, 1988. Green Book.

[Phi89]   Phillips and Sony Corporation. *System Description CD-ROM/XA*, 1989.

[Phi91]   Phillips and Sony Corporation. *System Description Recordable Compact Disc Systems*, 1991. Orange Book.

[Pic96]   R. W. Picard. A Society of Models for Video and Image Libraries. Technical Report 360, MIT Media Lab Perceptual Computing Section, 1996.

[PL90]   T. Pavlidis and Y.-T. Liow. "Integrating region growing and edge detection." *IEEE Transactions on Pattern Analysis and Machine Intelligence*, PAMI-12(3):225–233, March 1990.

[PM93]   W. B. Pennebaker and J. L. Mitchell. *JPEG Still Image Data Compression*. Van Nostrand Reinhold, New York, 1993.

[PM95]   R. W. Picard and T. P. Minka. "Vision texture for annotation." *Multimedia Systems*, 3:3–14, 1995.

[PMJA88] W. B. Pennbaker, J. L. Mitchell, G. Langdon Jr., and R. B. Arps. "An overview of the basic principles of the Q-Coder binary arithmetic coder." *IBM Journal of Research Development*, 32(6):717–726, November 1988.

[PS86] P. P-S. "The compact disc ROM: How it works." *IEEE Spectrum*, 23(4):44–49, Apr. 1986.

[PZM96] G. Pass, R. Zabih, and J. Miller. "Comparing images using color coherence vectors." In *ACM Conference on Multimedia*, Boston, MA, 1996. ACM.

[RB93] T. R. Reed and J. M. H. Du Buf. "A review of recent texture segmentation and feature extraction techniques." *CVGIP: Image Understanding*, 57:359–372, May 1993.

[RBE94] L. A. Rowe, J. S. Boreczky, and C. A. Eads. "Indexes for user access to large video databases." *Symposium on Electronic Imaging: Science and Technology*, February 1994.

[RDS76] L.R. Rabiner, J.J. Dubnowski, and R.W. Schafer. "Realtime digital hardware pitch detector." *IEEE Transactions on Acoustics, Speech and Signal Processing*, 24(1):2–8, Feb. 1976.

[Rei82] C.W. Reinholds. "Computer animation with scripts and actors." In *SIGGRAPH 82*, pages 289–296, 1982.

[Ril89] M. D. Riley. *Speech Time-Frequency Representation*. Kluwer Academic Publishers, Boston, 1989.

[Rip89] G. D. Ripley. "DVI—A digital multimedia technology." *Communications of the ACM*, 32(7):811–822, July 1989.

[RJ91] M. Rabbani and P. Jones. "Digital image compression techniques." In *Tutorial Texts in Optical Engineering*, Volume TT7. SPIE Press, 1991.

[RJ93] L. Rabiner and B.-H. Juang. *Fundamentals of Speech Recognition*. Signal processing. Prentice Hall, Englewood Cliffs, New Jersey, 1993.

[RLE97] Ch. Kuhmünch, R. Lienhart, and W. Effelsberg. "On the detection and recognition of television commercials." In *Proc. IEEE Conference on Multimedia Computing and Systems*, pages 509–516, Ottawa, Canada, June 1997.

[Ror93] M.E. Rorvig. "A method for automatically abstracting visual documents." *Journal of the American Society for Information Science*, 44, 1993.

[RSK92]   L.Q. Ruan, S.W. Smoliar, and A. Kankanhalli. "An analysis of low-resolution segmentation techniques for animate video." In *Proceedings of International Conference on Control, Automation, Robotics, and Computer Vision (ICARCV)*, pages 16.3.1–16.3.5, Bordeaux France, 1992.

[RSSS90] J. Rückert, B. Schöner, R. Steinmetz, and H. Schmutz. "A distributed multimedia environment for advanced CSCW applications." In *IEEE Multimedia '90, Bordeaux, France*, November 1990.

[Rus94]   G. Ruske. *Automatische Spracherkennung*. Oldenburg Verlag, 1994.

[SC95]   M.A. Smith and M. Christel. "Automating the creation of a digital video library." In *Proceedings of Third ACM International Conference on Multimedia*, pages 357–358, San Francisco, CA, November 1995.

[Sch92]   M. E. H. Schouten, editor. *The Auditory Processing of Speech: From Sounds to Words*. Mounton de Gouyter, Berlin, New York, 1992.

[Sch97a] J. Schönhut. *Document Imaging*. Springer-Verlag, Heidelberg, 1997.

[Sch97b] R. Schulmeister. *Grundlagen hypermedialer Lernsysteme*. Oldenbourg Verlag, München, Wien, 1997.

[SGC90]   S. Singhal, D. Le Gall, and C.-T. Chen. "Source coding of speech and video signals." *Proceedings of the IEEE*, 78(7), July 1990.

[SH86]   N. Suehiro and M. Hatori. "Fast algorithms for the DFT and other sinusoidal transforms." *IEEE Transactions on Acoustics, Speech and Signal Processing*, ASSP-34(3):642–644, June 1986.

[SH91]   R. Steinmetz and R. G. Herrtwich. "Integrierte verteilte Multimedia-Systeme." *Informatik Spektrum*, 14(5):280–282, October 1991.

[SHRS90] R. Steinmetz, R. Heite, J. Rückert, and B. Schöner. "Compound multimedia objects—Integration into network and operating systems." In *International Workshop on Network and Operating System Support for Digital Audio and Video, International Computer Science Institute, Berkeley, CA, USA*, November 1990.

[Sin90]   A. Singh. "An estimation-theoretic framework for image-flow computation." In *Proceedings of the Third International Conference on Computer Vision (ICCV)*, pages 168–177, Osaka, Japan, 1990.

[Sin92]   A. Singh. "Optic flow computation: A unified perspective." *IEEE Computer Society Press*, 1992.

[SL95]   S. N. Srihari and S. W. Lam. "Character Recognition." Technical Report CEDAR-TR-95-1, State University of New York at Buffalo CEDAR, January 1995.

[SPI94]   SPIE. Symposium on Electronic Imaging. In *Conference on Digital Video Compression on Personal Computers: Algorithms and Technologies*, Volume 2187, February 1994. SPIE/IS&T.

[SRR90]   R. Steinmetz, J. Rückert, and W. Racke. "Multimedia-Systeme." *Informatik Spektrum*, 13(5):280–282, 1990.

[Ste83]   G. Stern. "Bbop—A system for 3D keyframe figure animation." In *SIGGRAPH 83*, pages 240 –243, New York, July 1983. Introduction to Computer Animation, Course Notes 7.

[Ste94]   G. A. Stephen. *String Search Algorithms*. World Scientific Publishing Co. Pte. Ltd., 1994.

[Ste00]   R. Steinmetz. *Multimedia-Technologie: Grundlagen, Komponenten und Systeme*. Springer-Verlag, Heidelberg, October 2000. 3. Auflage (erstmalig mit CD).

[Sto88]   J. A. Storer. *Data Compression Methods and Theory*. Computer Science Press, 1988.

[Str88]   P. Strauss. Bags: The Brown Animation Generation System. Ph.D. Thesis CS-88-22, Computer Science Department, Brown University, Providence, RI, May 1988.

[Sut63]   I. E. Sutherland. "Sketchpad: A man-machine graphical communication system." In *SJCC, Spartan Books*, Baltimore, MD, 1963.

[Sv91]   F. Sijstermans and J. van der Meer. "CD-I full-motion video encoding on a parallel computer." *Communications of the ACM*, 34(4):81–91, April 1991.

[TATS94]   Y. Tonomura, A. Akutsu, Y. Taniguchi, and G. Suzuki. "Structured video computing." *IEEE Multimedia*, 1(3):34–43, 1994.

[Tec89]   Technical Manual, NeXT, Inc. *NeXT 0.9 Technical Documentation: Concepts*, 1989.

[Ton91]   Y. Tonomura. "Video handling based on structured information for hypermedia systems." In *Proceedings of International Conference on Multimedia Information Systems*, pages 333–344. Singapore, 1991.

[UGVT88] S. Uras, F. Girosi, F. Verri, and V. Torre. "A computational approach to motion perception." *Biological Cybernetics*, 60:79–97, 1988.

[Vet85] M. Vetterli. "Fast 2-D discrete cosine transform." In *Proceedings of IEEE ICASSP-85*, pages 1538–1541, Tampa, Florida, March 1985.

[VG91] E. Viscito and C. Gonzales. "A video compression algorithm with adaptive bit allocation and quantization." In *Proceedings of SPIE Visual Communications and Image Processing*, Volume 1605 205, Boston, MA, November 1991.

[VN84] M. Vetterli and H.J. Nussbaumer. "Simple FFT and DCT algorithms with reduced number of operations." *Signal Processing*, August 1984.

[Vor89] M. Vorländer. "Simulation of transient and steady-state sound propagation in rooms using a new combined ray-tracing/image-source algorithm." *Journal of the Acoustical Society of America*, 86(7):172–178, 1989.

[VS91] L. Vincent and P. Soille. "Watersheds in digital spaces: An efficient algorithm based on immersion simulations." *IEEE Transactions on Pattern Analysis and Machine Intelligence*, PAMI-13(6):583–598, June 1991.

[vZ89] B. A. G. van Luyt and L. E. Zegers. "The compact disc interactive system." *Phillips Technical review*, 44(11/12):326–333, November 1989.

[Wai88] A. Waibel. *Prosody and Speech Recognition*. Pitman, London, and Morgan Kaufmann Publishers, San Mateo, CA, 1988.

[Wal91] G. K. Wallace. "The JPEG Still Picture Compression Standard." *Communications of the ACM*, 34(4):30–44, April 1991.

[WC72] H.R. Wilson and J.D. Cowan. "Excitatory and inhibitory interactions in localized populations of model neurons." *Biophysics Journal*, pages 1–24, 1972.

[Wep92] Artur Wepner. CD Software for the CD-WO Writer. *Personal Communication*, 1992.

[Wil89] C. J. Williams. "Creating multimedia CD-ROM." In *Proceedings of 7th Conference on Interactive Instructions Delivery*, pages 88–92, March 1989.

[Wol90] G. Wolberg. *Digital Image Warping*. IEEE Computer Society Press Monograph, 1990.

[WSF92] H. E. Wolf, F. Strecker, and G. Fries. Text-Sprache-Umsetzungen für Anwendungen bei automatischen Informations- und Transaktionssystemen. In H. Mangold, editor, *Sprachliche Mensch-Maschine-Kommunikation*. Oldenbourg, München, Wien, 1992.

[WVP88] G. Wallace, R. Vivian, and H. Poulsen. "Subjective testing results for still picture compression algorithms for international standardization." In *IEEE Global Telecommunications Conference*, pages 1022–1027, November 1988.

[WWB88] A.M. Waxman, J. Wu, and F. Bergholm. "Contour evolution, neighbourhood deformation, and global image flow." In *Proceedings of IEEE Conference on Computer Vision and Pattern Recognition*, pages 717–723, Ann Arbor, 1988.

[YL95] M. Yeung and B. Liu. "Efficient matching and clustering of video shots." In *IEEE International Conference on Image Processing (ICIP), Special Session on Digital Libraries*, Washington, DC, October 1995.

[Zam89] P. Zamperoni. *Methoden der digitalen Bildverarbeitung*. Vieweg, Braunschweig, Wiesbaden, 1989.

[ZGST94] H.-J. Zhang, Y. Gong, S. W. Smoliar, and S. Y. Tan. "Automatic parsing of news video." In *Proceedings of IEEE Conference on Multimedia Computing and Systems*. IEEE, Boston, MA, May 1994.

[ZKS93] H.-J. Zhang, A. Kankanhalli, and S. W. Smoliar. "Automatic partitioning of full-motion video." *Multimedia Systems*, 1(1):10–28, January 1993.

[ZLB+87] T. Zimmerman, J. Lanier, C. Blanchard, S. Bryson, and Y. Harvill. "A hand gesture interface device." In *Proceedings of Computer Human Interaction and Graphics Interface (CHI+GI)*, pages 189–192, New York, NY, 1987. ACM.

[ZLSW95] H. J. Zhang, C. Y. Low, S. W. Smoliar, and J. H. Wu. "Video parsing, retrieval and browsing: An integrated and content-based solution." In *Third ACM Conference on Multimedia*, pages 15–24, San Francisco, 1995.

[ZS94] H.-J. Zhang and S. W. Smoliar. "Developing power tools for video indexing and retrieval." In *Proceedings SPIE Conference on Storage and Retrieval for Image and Video Database*, San Jose, CA, 1994. SPIE.

[ZSW+95] H. Zhang, S. W. Smoliar, J. H. Wu, C. Yong Low, and A. Kankanhalli. "A video database system for digital libraries." *Advance in Digital Libraries, Leisure Note in Computer Science*, 1995.

[ZW94]   R. Zabih and J. Woodfill. "Non-parametric local transforms for computing visual correspondence." In Jan-Olof Eklundh, editor, *3rd European Conference on Computer Vision*, Stockholm, Volume LNCS801, pages 151–158. Springer-Verlag, 1994.

[ZWLS95] H. Zhang, J.H. Wu, C.Y. Low, and S. W. Smoliar. "A video parsing, indexing and retrieval system." In *Proceedings of Third ACM International Conference on Multimedia*, pages 359–360, San Francisco, CA, November 1995.

Index

Numerics

16CIF, 137
4CIF, 137

A

AAL
    AAL 2, 161
    AAL5, 159
Absorption layer, 194
Access Unit Layer, 159, 162, 164
Access Units, 159
AC-coefficient, 128, 129, 130
Achromatic light, 76
Acoustic analysis, 39
Acoustic signals, 22
Acoustics
    doubling effect, 24
    frequency masking, 26
    Haas effect, 25
    masking effect, 26
    physical, 23, 24
Adaptation dynamics, 72

Adaptation Layer, 161
Adaptive Differential Pulse Code
    Modulation, 187
Adaptive DPCM, 119
Adaptive Pulse Code Modulation, 40
ADC, 27
Adobe
    Type1 format, 49
ADPCM, 40, 112, 119, 187, 203
Aliasing, 165
Allophones, 34
Amplitude, 26
    scaling, 150
Analog-to-digital converter, 27
Animation, 71, 95
    control, 98
    conventional, 95
Annotation, 209
Apple Macintosh Picture Format, 55
AQUIRE, 56
Area, 178
Arithmetic coding, 116
ASAS, 98
ASCII, 8

Aspect ratio, 80
ATM, 150, 151, 159, 161
AU layer, 162
Audio
    access units, 146
    amplitude, 26
    coding, 40, 144
        quality levels, 145
    compact disc, 169
    cut, 228
    cut detection, 226, 227
    data rate, 175
    indicators, 206
    samples, 26
    segmentation, 207
    signal, 107
    stream, 146
    technology, 21
Audio bit stream, 179
Audio coding, 144
Audio data stream, 179
Audio/Video Support System, 193
Audiovisual objects, 152
Authoring tools, 102
Autocorrelation method, 227
AVO, 155
AVSS, 193

B

B frame, 142, 144, 211
Bandpass filters, 212
BBOP system, 99
Bidirectionally predictive-coded
    pictures, 142
Binarization, 61
Bit rate
    constant, 16
    variable, 17
Bitmap, 45, 52, 107
Blanking interval, 92

Block, 181
    logical, 181
    physical, 181
Block layer, 147
Block matching, 233
Block vectors, 158
Blocks, 137, 178, 181
BMP, 52
    color model, 52
Book A-E, 198
Brightness, 231
    signal, 81
BRIM, 48
Burst error, 177
Byte stuffing, 113

C

CAD, 30
    model, 30
Camera effects, 206
Camera motion, 206, 215, 216, 218, 219
Camera parameters, 158
Canny operator, 62
CAPTAIN, 8
CAT, 56
Cathode Ray Tube, 79
CATV, 93
CAV, 174
CCD, 46
    camera, 46
    scanner, 46
CCIR, 90
CCIR 601, 107, 137, 148, 150
CCITT, 136
CCV, 57
CCVS signal, 88
CD, 172
CD Bridge Disc, 191, 192
CD publishing, 195

CD quality, 107
CD technology, 105
CD track, 175
CD-DA, 109, 144, 170, 176, 177, 181, 186
   capacity, 176
CD-I, 170, 185, 187, 188
   audio coding, 189
CD-MO, 170, 196, 204
CD-R, 194
CD-R/W, 204
CD-ROM, 169, 170, 180, 186, 203
CD-ROM mode 1, 182, 185, 186, 192, 193, 203
   capacity, 182
   data rate, 182
CD-ROM mode 2, 183, 186, 192, 203
   capacity, 183
   data rate, 183
CD-ROM technologies, 188
CD-ROM/XA, 170, 185, 186, 189, 203
CD-RTOS, 189
CD-RW, 171, 197
CDTV, 193
CD-WO, 170, 192
Census transform, 219
Center clipping, 227
Central projection equation, 46, 74
Cepstrum technique, 227
CEPT, 8
CGA, 86
CGEG, 120
CGM, 55
Channel bit stream, 179
Channel bits, 176, 177
Channel vocoder, 41
Character recognition, 208
Character segementation, 208
Chromatic scaling, 215, 220
Chrominance, 19, 81, 83, 84, 87, 107, 141, 158, 190, 193
   difference signal, 137
   signal, 88, 91
CIF, 137, 150
Classification, 208, 229
   structural, 208
Classification problem, 229
Clock, 176
Clusters, 230
CLUT, 86, 190
CLV, 175
CMYK, 50
Coarticulation, 35
Codec, 136
Coding, 90, 141
   arithmetic, 110, 115, 116, 131, 133
   as vector, 118
   asymmetric, 112, 165
   bi-level, 120
   component, 90, 91, 93
   composite, 90, 93
   diatomatic, 114
   DPCM, 143
   efficiency, 154
   entropy, 110
   hierarchical, 121
   Huffman, 110, 115, 131
   hybrid, 110, 154
   in MPEG-4, 156
   integrated, 91
   interframe, 111
   intraframe, 112, 141
   lossless, 110, 121
   lossy, 110
   of P frames, 143
   progressive, 132
   relative, 117
   requirements, 106
   run-length, 110, 113
   shape, 157

source, 110, 111
statistical, 114
subband, 110, 117
symmetric, 112
table, 118
transformation, 117
with metadata, 165
YUV, 122
Color, 57, 207
    channel, 48
    coherence vector, 57
    crosstalk, 83
    depth, 53
    encoding, 48
    histogram, 57
    information, 81, 83, 84
    look-up table, 86, 97, 120, 190
    model, 50
    perception, 209
    subsampling, 140
    table, 53
    values, 212
Color Graphics Adapter, 86
Color Look-Up Table, 86, 97, 120, 190
Comb filter, 83
Commodore Dynamic Total Vision, 193
Common Intermediate Format, 137
Compact Disc, 172
    Bridge Disc, 191
    Digital Audio, 144, 170, 175
    Interactive, 170, 185, 188
    Interactive Ready Format, 190, 191
    Magneto Optical, 170, 196
    Read Only Memory, 170, 180
    Read Only Memory Extended
        Architecture, 170
    Recordable, 194
    Write Once, 170
Compression
    adaptive, 118

asymmetric, 140
fractal, 165
symmetric, 108, 140
wavelet, 156
Compression techniques
    adaptive, 118
Computer graphics, 55
Computer Graphics Metafile, 55
Computer tomography, 73
Computer vision, 57
Computerized Axial Tomography, 56
Connected component analysis, 208
Constant Angular Velocity, 174
Constant Linear Velocity, 175
Content analysis, 229
Continuous-tone approximation, 76
Contraction factor, 166
Contractivity condition, 166
Control byte, 178
Convolution, 60
    kernel, 60
    mask, 60
Correspondence, 225
Cosine transform, 111
Cross Interleaved Reed-Solomon Code,
    178, 180
Crosstalk, 91, 173, 199
CRT, 77, 79
Cut, 206, 214, 217, 219
Cut boundaries, 220
Cut detection, 205, 215
    audio, 227
    pixel-based, 215
Cuts, 214

D

D frame, 142, 144
DAC, 27
DAT, 144, 169
Data bit stream, 179

Data compression, 105
Data glove, 53, 99
Data stream, 16, 139, 146
    constant bit rate, 16
    granularity, 20
    interrelated, 18
    irregular, 17
    non-interrelated, 18
    strongly regular, 16
    variable bit rate, 17
    weakly regular, 16
Data streams, 7, 13, 15
    aperiodic, 15
    strongly periodic, 15
    weakly periodic, 15
DC-coded pictures, 142
DC-coefficient, 128, 130
DCT, 110, 111, 117, 120, 122, 124, 127, 132, 143, 144, 145, 157, 166, 221
    AC-coefficient, 128
    coefficients, 215
    DC-coefficient, 128
    inverse, 128
    shape adaptive, 157
    zig-zag sequence, 130
Deblocking filter, 139
Decoder, 189
Decryption, 201
Delay
    end-to-end, 108, 150
Delay jitter, 14
Delivery Multimedia Integration Framework, 163
Delta function, 73
Delta modulation, 112, 118
Demultiplexing, 159, 161
Depth perception, 81
Descrambling, 201
DFT, 127
DIAL, 98
Dialogue mode, 108, 109
Diatomic encoding, 114
Differential Pulse Code Modulation, 40, 118
Digital Audio Tape, 144, 169
Digital speech processing, 32
Digital TeleVision, 93
Digital Versatile Disc, 198
Digital Video Broadcasting, 87
Digital Video Interactive, 170, 185, 193
Digital-to-analog converter, 27
Digitization, 90
Diphone, 35
Directory tree, 183
Discrete Cosine Transform, 117, 120, 127
Display
    controller, 75
Display byte, 178
Dissolve, 206, 214, 218, 219, 220
Distance function, 222
Distance measure, 231
    Euclidean, 231
Dithering, 76
    matrix, 77
    monochrome, 77
DM, 110
DMIF, 161, 163
Double buffering, 97, 100
Double-layer DVD, 199
Doubling effect, 24
DPCM, 40, 110, 118, 119, 138, 143, 159
DTVB, 93
Dual channel, 146
DVB, 87, 93
    Guidelines Document, 93
DVB-TXT, 93
DVD, 169

Audio, 198
Consortium, 171, 198
ROM, 198
standard, 198
Video, 198
DVD technology, 105
DVD-RAM, 204
DVI, 170, 185, 193
DXF, 55
Dynamics, 100

E

Echo processes, 30
ECMA, 109, 170
ECMA standard, 180
Edge
    extraction, 60, 218
    image, 207, 218
    pixel, 218
Edges, 60, 206
Editing effects, 206, 214, 220
    chromatic scaling, 215
    edge extraction, 215
    histogram changes, 215
Editing images, 220
EGA, 86
Eight-to-Fourteen Modulation, 176, 180, 201
Eight-to-Fourteen+ Modulation, 201
El Torito Extension, 184
Electronic publishing, 71
Elementary Stream
    interface, 162
Encoding
    diatomatic, 114
    differential, 117
    run-length, 53
Encryption systems, 94
End-to-end delay, 13, 150, 151
Enhanced Graphics Adapter, 86

Entropy coding, 110, 112, 133
Entropy encoding, 130, 134, 143
EPS, 49
Error
    correction code, 112
    handling, 177
Euclidean distance, 58
Euler-Lagrange equations, 100
Event, 97
Eye
    motion resolution, 81

F

Fade rate, 220
Fade-in, 214, 219, 220
    from black, 220
Fade-out, 206, 214, 217, 218, 219, 220
    to black, 220
Fan-beam projection, 73
Fast forward, 109
Fast Fourier Transform, 145
Fast offset, 228, 229
Fast onset, 228, 229
Fast onset filter, 229
FDCT, 127
Feature vector, 209, 224, 229
FFT, 110, 117, 145
Field of view
    angular, 80
File format
    logical, 183
Filler bit, 179
Film, 82
Film genre, 206, 229
Filter
    comb, 83
    deblocking, 139
    notch, 83
First wave-front law, 25
FlexMux Layer, 159, 162

Flicker effect, 82, 83, 86
Focal length, 74
Font
    scalable, 49
Form 1, 186
Form 2, 186
Formant synthesis, 36
Formant tracking, 111
Formants, 36, 111
Forward DCT, 127
Forward error correction, 155
Fourier transform, 73, 117, 226
    1D, 73
    2D, 73
    fast, 117
Fractal, 52
Frame, 81, 177, 178
    buffer, 86, 96
    grabber, 46
    rate, 85, 109
    size, 109
Frames, 146, 210
Frequency
    bands, 117
    distribution, 206, 226
    domain, 111
    masking, 26
    transformation, 145
    transitions, 226
Full picture, 82
Fundamental frequency, 226, 227, 229

G

G.700, 120
G.721, 120
G.728, 120
GENESYS, 98
GIF, 50
GIF24, 50
GIF87a, 50
GIF89a, 50
GKS, 8, 55
Gradient, 61
    orientation, 61
Graphic
    application model, 54
    application program, 54
    interactive system, 54
    primitives, 54
Graphical Kernel System, 55
Graphics, 45, 53
    attributes, 45
    computer-assisted, 55
    dynamic, 72
    library, 75
    meta file format, 55
    output, 75
    static, 72
    storing, 54
    synthesis, 55
    system, 54
Graphics Interchange Format, 50
Gray value, 206, 212
Gray-level co-occurrence matrix, 59
Green Book, 170, 188
Group of pictures
    layer, 147
GSM, 152, 167

H

H.223, 139, 161
H.261, 108, 135, 136, 140, 167, 203, 211
    group of blocks, 137
    interframe, 138
    intraframe, 137
    loop filter, 136
    macro block, 137
    motion vector, 138
    quantization, 139

time out, 139
H.263, 105, 108, 136, 167, 211
    advanced prediction mode, 138
    arithmetic coding, 138
    Overlapped Block Motion Compensation, 138
H.263+, 139
H.263L, 139
H.320, 136
Haar transform, 117
Haas effect, 25
Hadamard transform, 117
Half picture, 82
Hausdorff distance, 219
HDTV, 88, 93, 141, 148, 152
    High 1440 Level, 89
    High Level, 89
    standard, 89
Head-related transfer function, 24, 30
Hewlett Packard Graphics Language, 55
High Definition Television, 88
High Sierra Proposal, 170
Histogram, 62
Homogeneity
    condition, 63
    criterion, 64
Hough transform, 62
HPGL, 55
HRTF, 24, 30
Huffman, 145
    coding, 115, 131
Hybrid
    CD-R, 195
    coding, 110
Hyperlink, 102

I

I frame, 141, 143
Icon
    XBM, 51
    XPM, 51
IDCT, 128
IGES, 55
IID, 23
Image, 45
    analysis, 56
        conditioning, 69
        extraction, 70
        grouping, 70
        marking, 70
        matching, 71
    binary, 47
    capturing, 48
    compression
        fractal, 105
    enhancement, 208
    formats, 48
    generation, 56
    improvement, 56
    monochromatic, 47
    output, 75
    preparation, 134, 137, 140
    processing, 127, 134, 141, 143
    projecting, 72
    recognition, 56, 66, 69
    reconstructing, 72
    refresh frequency, 141
    segmentation, 61
        edge-oriented, 62
        pixel-oriented, 62
        region-growing, 63
        region-oriented, 63
        split-and-merge, 64
        water-inflow, 65
        watershed method, 65
    similarity, 230
    spatial resolution, 48
    stereoscopy, 73
    storage, 48
    synthesis, 71

ImagePac, 193
Imitation, 30
Inbetween processing, 96
Index point, 179, 181, 191
Indicators, 206
    semantic, 206, 227, 229
    syntactic, 206, 226, 229
Information filtering, 205
Information unit, 7, 16, 19
Initial Graphics Exchange Standard, 55
Intellectual Property Rights, 156
Inter- and intraclass distance, 230
Interactive Video Disc, 174
Interaural intensity difference, 23
Interaural time difference, 23
Interclass distance, 230
Interference, 85
    elimination, 85
Interframe, 137
Interlaced scanning, 137
Interlaced video, 148
Interleaving, 124, 187
Inter-pixel interpolation, 74
Interpolation, 99, 113
    linear, 96
Intra coded pictures, 141
Intraframe, 137
Inverse DCT, 128
IP, 159, 161
IPR, 156
IS, 120
ISDN, 135, 151, 167
    B-channels, 135
    D-channel, 135
ISO, 109, 120
ISO 11172, 148
    data stream, 148
ISO 9660, 183, 190
    standard, 170
ISO Intermediate Symbol Sequence, 131
ITD, 23
ITU, 90, 109, 119

J

JBIG, 120
Jitter, 14, 15
Joint Photographic Experts Group, 120
Joint stereo, 145, 146
Joliet File System, 184
Joystick, 189
JPEG, 8, 105, 120, 140, 166, 203
    abbreviated format, 121
    arithmetic coding, 131
    base mode, 121, 122
    baseline process, 122, 127
    common context, 121
    expanded lossy, DCT-based mode, 122
    expanded, lossy DCT-based mode, 132
    hierarchical mode, 122, 135
    Huffman coding, 131
    image preparation, 122
    lossless mode, 122, 134, 166
    lossy, sequential DCT-based mode, 122, 126
    Minimum Coded Unit, 122
    modes, 121, 122
    non-interleaved image processing, 125
    requirements, 121

K

Kell factor, 80
Key frames, 95, 98, 220, 222
Key scenes, 205
Kinematics, 100

L

Label, 172
Land, 171, 175, 194, 196
Lands
    length, 176
Language, 227
LaserVision, 173
Layer
    reflective, 172
Layered coding, 133, 150
LDU, 19, 210
Lead-in area, 179, 194
Lead-out area, 179, 194
Lerping, 96
Light
    achromatic, 76
    quality, 76
Likelihood ratio, 215
Line fitting, 70
Linear interpolation, 96
Linear predictive coding, 36
Logical data unit, 19
Long-playing record, 169
Look-Up Table, 97
Low pass filter
    optical, 139
LP, 169
Luminance, 19, 81, 83, 84, 87, 91, 107, 140, 141, 158, 190, 192
    crosstalk, 83
    signal, 91, 137
LUT, 86
LZW, 50

M

Macro block, 137, 138, 141, 158
Macro block layer, 147
Mahalanobis distance, 210
Main Profile, 149

Markov random fields, 60
Masking effect
    frequency range, 26
    time range, 26
Matrix, 47
MATV, 93
Maximum likelihood estimation method, 227
MCU, 122, 125, 141
Media, 7
    computer-controlled, 12
    continuous, 10, 12, 107
    discrete, 10, 12
    independence, 11
    independent, 12
    information exchange, 9
    overflow, 174
    perception, 8
    presentation, 8
        dimensions, 10
        spaces, 9
        values, 9
    representation, 8, 11
    storage, 9
    time dependent, 10
    time-independent, 10
    transmission, 9
Medium, 7
Mesh, 156
Metadata, 165, 205
    track, 205
Metric
    L1, 210
    L2, 210
Microwave, 93
MIDI, 21, 30
    cable, 31
    clock, 32
    data format, 31
    devices, 31

Index

interface, 32
message, 31
noise effects, 32
port, 31
sequencer, 32
synthesizer, 31
time code, 32
Minimum Coded Unit, 125, 126
Mixed mode disc, 180, 186, 188, 191
MMDS, 93
Mode
    hierarchical, 135
Morpheme, 34
Motion, 210
    analysis, 233
    compensation, 118, 136, 157, 167, 218, 219
    continuity, 81
    resolution, 81
    vectors, 111, 138, 206, 210
        block oriented, 211
        pixel oriented, 211, 212
Motion compensation, 167
Motion dynamics, 95
Motion JPEG, 167, 221
Movement
    compensation, 89
Movement dynamics, 72
Moving Worlds, 102
MPEG, 105, 108, 120, 139, 167, 190, 201, 203
    audio, 187
    audio access unit, 146
    B frame, 142, 144
    channels, 145
    coding, 141
    D frame, 142, 144
    group of pictures, 144
    I frame, 141, 143
    image preparation, 140
    image processing, 141
    image refresh frequency, 141
    layer, 145
    macro block, 141
    motion vector, 143
    multiplexing, 147
    P frame, 141, 143
    prediction, 141
    psychoacoustic model, 145
    quantization, 144
    quantization characteristic curve, 143
    search range, 143
    sequence layer, 146
    slot, 146
    stereo sound, 145
    subbands, 145
    System Decoder, 201
    system definition, 140, 147
MPEG-1, 108, 211, 221, 233
MPEG-2, 148, 167, 211, 221, 233
    audio coding, 150
    basic multiplexing, 151
    High Profile, 149
    Main Profile, 149
    Program Stream, 151
    scaling, 150
    system, 151
    Transport Stream, 151
MPEG-3, 152
MPEG-4, 139, 152, 167
    Access Unit, 159
    Access Unit Layer, 161, 162
    AL-PDU header, 164
    audiovisual objects, 152
    camera parameters, 158
    composition units, 164
    compression efficiency, 154
    content-based access, 154
    content-based scalability, 153

Elementary Streams, 159
error protection, 157
error robustness, 155
layers, 156
macro blocks, 158
multiplexing sublayer, 161
Pred 1 - Pred 3, 158
protection sublayer, 161
real-time data, 164
session, 164
streams, 159
time bases, 164
TransMux, 159
MPEG-7, 165
consistency checking, 165
display strategies, 165
searching techniques, 165
Multi-channel sound technique, 30
Multilingual, 150
Multimedia, 7
system, 11
Multiplexing, 159
technique, 107
Multi-scale simultaneous autoregression, 60
Multiview video, 153
Music, 30
Musical Instrument Digital Interface, 21, 31

N

National Television Systems Committee, 87
Nearest neighbor, 230, 232
Noise, 22
Notch filter, 83
NTSC, 83, 87
standard, 82
Nyquist Theorem, 27, 90

O

Object
category, 67
class, 67
extraction, 62
movement, 216
orientation, 67
position, 67
recognition, 56, 66
segmentation, 63
Objects, 207
OBMC, 138
OCR, 56, 233
systems, 207
Office automation, 71
Offset, 226, 228
Onset, 226, 228
fast onset, 229
Optical Character Recognition, 46, 56, 207
Optical flow, 212
aperture problem, 213
correlation between regions, 212
correlation-based methods, 212
deformable bodies, 213
differential methods, 212
energy-based methods, 212
methods, 212
phase-based methods, 212
physical correspondence, 213
Optical low pass filter, 139
Orange Book, 171, 194, 196
Overlapped Block Motion Compensation, 138

P

P frame, 141, 143, 211
coding, 143
Packaging, 174

Packetized Elementary Stream (PES), 151
Padding, 162
PAL, 83, 88, 91
    standard, 82, 107
Pan, 96
Parallel-beam projection, 73
Parametric systems, 41
Partial scene, 220
Path table, 183
Pattern
    discovery, 56
    recognition, 56
        dynamic, 56
        static, 56
    substitution, 114
Pay-TV
    services, 94
PBM, 52
PBMplus, 52
PCM, 8, 15, 40, 91, 118, 145, 175, 202
    quantization, 47
PDA, 207
PDU, 19
Pencil test, 96
Performance measures, 230
    efficiency, 231
    inter- and intraclass distance, 230
    nearest neighbor, 230
Personal Digital Assistant, 207
PGM, 52
Phase Alternating Line, 88
Phase errors, 88
PHIGS, 55
Phoneme, 34, 35
Phonetic analysis, 39
Photo CD, 170, 192
PICT, 55
Picture
    preparation, 111
    processing, 111
    refresh rate, 82
Picture layer, 147
Pit, 171, 175, 194, 196, 200
Pitch, 227
Pits
    length, 176
Pixel, 45, 47, 80, 111, 123
    matrix, 76
Pixmap, 55
PNG, 50
PNM, 52
Portable Anymap, 52
Portable Bitmap, 52
Portable Bitmap plus, 52
Portable Graymap, 52
Portable Pixmap, 52
PostScript, 49
    encapsulated, 49
    Level 1, 49
    Level 2, 49
    Level 3, 49
PPM, 52
Prediction, 117, 142
    bidirectional, 144
Predictive-coded pictures, 141
Predictors, 134
Premastered area, 196
Preparation, 111
Processing, 111
Profile, 231
Program area, 179
Program stream, 93
Programmer's Hierarchical Interactive Graphics System, 55
Projection, 212
    disk theorem, 73
    fan-beam, 73
    parallel-beam, 73
Proportionality constant, 100

Prosody, 35
Protective layer, 172
Protocol data unit, 19
Proximity sensor, 103
Psychoacoustics, 23
    first wave-front law, 25
Pulse Code Modulation, 15, 40
Pulse response, 29

Q

QCIF, 137
QoS, 161
Quadrature Amplitude Modulation, 88
Quadrophony, 9
Quality of Service, 159
Quantization, 27, 52, 85, 90, 111, 129, 134, 143, 230
    16-bit linear, 175
    characteristic curve, 118, 143
    steps, 90
    vector, 165
Quantizer, 133
Quarter CIF, 137
QuickTime, 105

R

Radon transform, 73
Raster, 76
    display, 75, 85
    graphics, 76
    lines, 76
Read Only Memory, 174
Real-time
    operation, 164
Red Book, 170, 179, 180, 185, 188
Reed-Solomon, 177
Refresh buffer, 75
Region-growing, 63
Regular CD-R, 195

Relative coding, 117
Resequencing, 225
Resolution, 85
Resource Interchange File Format, 48
Retinal beam, 74
Retrieval mode, 108, 109
Revision, 175
Rewind, 109
RGB, 81, 113, 122
    color, 206, 230
    mode, 190
    signal, 83
    triple, 48
RIFF, 48
Rockridge Extensions, 184
Rotation, 98, 158
Rotation animation, 100
Rotation delay, 185
Rotoscoping, 99
RTP, 159
Run-length coding, 113
Run-length encoding, 53, 131, 190

S

SA-DCT algorithm, 158
Sampling, 90
    frequency, 90
    interlaced, 80
    progressive, 80
    rate, 82
Sampling rate, 27
Scalability, 153
Scaling, 98
Scefo, 98
Scene, 210
    analysis, 57
    length, 229
    segmentation, 158
Scientific visualization, 71
S-Dynamics System, 98

SECAM, 87
Seed pixel, 63
Seek time, 185
Segmentation, 206
    scene, 158
Self-similarity, 165
Semantic analysis, 39
Semantics, 39
Sequence layer, 146
Sequential Couleur avec Memoire, 87
Sequential forward selection, 231
Session, 195
Shape coding, 157
Shot, 210, 214
Showscan technology, 81
SIF, 92
Signal feedback algorithm, 30
Signal-to-noise ratio, 176
Silence suppression, 120
Similarity
    between images, 207
Simple Raster Graphics Package, 55
Simulation, 30, 71
Sink, 13
Sketchpad, 99
Slant transform, 117
Slice, 147
Slope overload, 119
Slots, 146
SMPTE, 32, 148
Sobel operator, 60
Sound, 21
    amplitude, 23
    basic frequency, 41
    concatenation
        in frequency range, 36
        in time range, 34
    decibel, 25
    direct path, 29
    dispersion, 28
    path, 28
    doubling effect, 24
    effects, 28
    Fletcher-Munson graphics, 25
    frequency, 22
    Haas effect, 25
    loudness, 25
    noise, 22
    non-periodic, 22
    perception, 23
    periodic, 22
    periodicity, 22
    physical acoustic perspective, 23
    pressure level, 25
    psychoacoustic perspective, 25
    psychoacoustics, 23
    spatial, 28
    speech, 22
    stereo, 28
    stimulus, 29
    three-dimensional projection, 28
    two-channel, 28
    unvoiced, 41
    voiced, 41
    wave form, 21
Sounds
    explosive, 36
    fricative, 36
Source, 13
    coding, 110, 111
    encoding, 41
    Input Format, 92
Spaceball, 53
Spatial
    domain, 58
    frequency domain, 58
    integration, 76
Spectral selection, 133
Speech, 22, 32, 107
    allophones, 34

    basic frequency, 34
    carrier words, 35
    coarticulation, 35, 36, 37
    digital processing, 32
    fundamental frequency, 206, 226, 227
    generation, 33
    input, 37
        speaker-dependent, 40
        speaker-independent, 40
    morpheme, 34
    output, 33
    phoneme, 34
    playout, 34
    pronunciation, 37
    prosody, 35, 36, 37
    recognition, 33, 38, 207
    semantics, 39
    signals, 32
    sound concatenation, 34
    synthesis, 33, 36
    transmission, 40
    tube models, 36
    unvoiced sound, 34, 36
    voiced signals, 33
    voiced sound, 34, 36
SPL, 25
Spline, 96
Split-and-merge, 64, 233
SQCIF, 137
SRGP, 55
Standard
    de facto, 109, 181, 203
    de jure, 109
Statistical coding, 114
Stereo, 28, 146
Stereo sound, 145
Stereophony, 9
    binaural, 30
Stereoscopic parallax, 74

Stereoscopy, 73, 74
Storage capacity, 105
Stream Multiplex Interface, 161
String search, 224
Subband coding, 117
Subband encoding, 41
Subbands, 145
Subchannel, 178
    bit, 178
Subheader, 186
Subsampling, 92, 137
Substrate, 172
    layer, 171
Subtitles, 208
Successive approximation, 133
Super Video Graphics Array, 87
Surround, 150
SVGA, 87
Synchronization
    audio and video, 109
    pattern, 178
    time, 184
Synchronization bit, 179
Syntactic analysis, 39
System
    area, 183
    decoder, 161
    definition, 140, 147

T

Table of contents, 192, 194
Tagged Image File Format, 50
Teletext, 93
Television
    data rate, 107
    high-resolution, 107
Template matching, 71, 208
Templates, 71
Terminal symbols, 114
Text

Index

recognition, 233
representation, 106
Texture, 52, 58, 207
    analysis
        statistical, 58
        structural, 58
    coarseness, 58
    contrast, 58
    homogeneity, 59
    information, 158
    local gray-level variance, 58
    orientation, 58
    regularity, 58
Texturing algorithms, 158
ThingLab, 99
Thresholding, 70
TIFF, 50
    baseline, 50
    color model, 50
    extensions, 50
    PackBits compression, 50
Track, 173, 178, 179, 180, 181
    density, 200
    pregap, 180, 191
Trackball, 53
Trailer film, 96
Transcoding, 154
Transcription, 36
    phonetic, 36
Transform
    discrete cosine, 117, 120, 127
    fast Fourier, 117, 145
    forward DCT, 127
    Fourier, 117
    Haar, 117
    Hadamard, 117
    Slant, 117
    Walsh, 117
Transformation, 68
    coding, 117

Translation, 98, 158
Transmission mode
    asynchronous, 13
    isochronous, 14
    synchronous, 14
TransMux
    Channel, 161
    Layer, 159, 161
Transport stream, 93
Tree-structured wavelet transform, 60

U

UDP, 159, 161
Update dynamics, 95

V

Variance-covariance matrix, 210
Vector
    field, 212
    images, 106
    quantization, 114, 165
VGA, 47, 86
Video, 79
    accelerator chips, 87
    buffer verifier, 146
    controller, 75, 86
    digitalization, 90
    disc, 173
    encoding, 140
    indexing systems, 214
    indicators, 206
    sequence, 107
    stream, 146
    telephony, 167
    track, 220
        black, 220
Video CoDec for Audiovisual Services at px64 Kbits/s, 136
Video for Windows, 105

Video Graphics Array, 47, 86
Video Long Play, 170, 173
Videophone, 148
Viewing distance, 80
Virtual Reality, 101, 154
    Modeling Language, 101
VLP, 170
Vocoder, 41
Voice fundamental frequency, 111
Volume, 206, 226
Volume descriptor, 183, 184
    primary, 183
    supplementary, 183
Vowels, 227
VRML, 101, 156

W

Walsh transform, 117
Warping, 219
Water-inflow
    hierarchy graph, 66
Wavelet compression, 156
Wavelets, 211
Width-to-height ratio, 80
Windows Metafile, 55
Wipe, 206, 214, 218, 219
WMF, 55
Word models, 39
World coordinate, 74

WORM, 170, 174, 194
Write Once Read Many, 174

X

X11
    Bitmap, 51
    Pixmap, 51
XBM, 51
XPM, 51
    hot spot, 51

Y

Yellow Book, 170, 188
YIQ
    signal, 84, 122
YUV, 113
    coding, 122
    color images, 123
    format, 140
    mode, 190
    signal, 84, 122

Z

Zero suppression, 113
Zig-zag sequence, 130
Zoom, 96, 158, 193, 206
    center, 206

About the Authors

RALF STEINMETZ is Professor of Multimedia Communications at the Technical University of Darmstadt, Chairman of the Board of the Hessian Telemedia Technology Competence Center httc, and a leader in the worldwide community of advanced multimedia researchers. He initiated the *IEEE Multimedia Magazine,* has been editor of various journals, is IEEE Fellow, and also recently became an ACM Fellow. His textbook, *Multimedia Technology: Principles, Components and Systems* has been the authoritative resource in various languages for American and European universities since its first publication in 1995.

KLARA NAHRSTEDT is Professor, Department of Computer Science, University of Illinois at Urbana-Champaign. Her research interests include Quality of Service provisioning for real-time multimedia processing and communication systems. She is the main editor of the *ACM/Springer Multimedia Journal* and has served on the editorial board of *IEEE Transactions on Multimedia,Computer Networks Journal,* and *Journal on Multimedia Applications and Tools*.

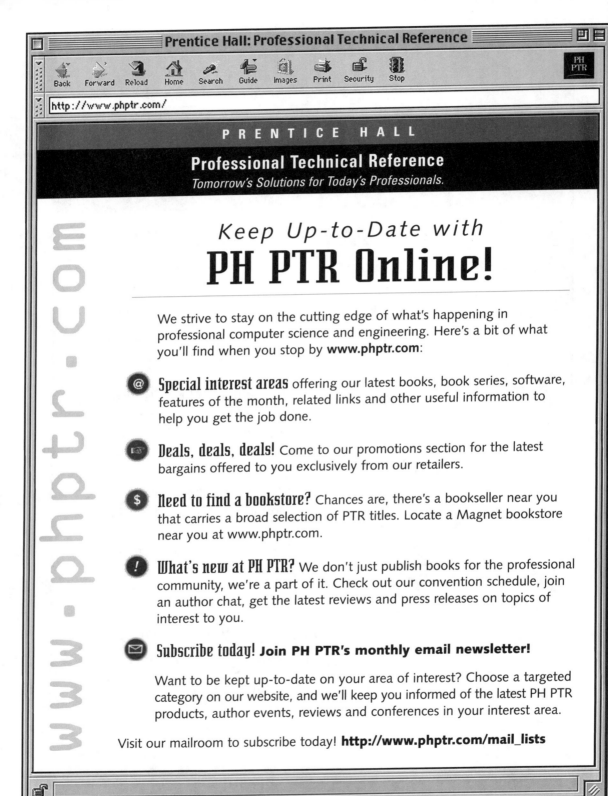